SRA MATH SKILLBUILDER

orange

McGraw Hill SRA

Columbus, OH

Photo Credits

22, Gary Conner/PhotoEdit; **36,** Robert Cerri/Corbis-Stock Market; **48, (l)** Daniel J. Cox/Tony Stone Images; **(r)** Kennan Ward/ Corbis Stock Market; **64,** Doug Martin; **74,** Sante Fe Southern Railroad; **92,** David Young-Wolff/PhotoEdit; **106 (l),** Tony Freeman/ PhotoEdit; **(r),** Jane Sapinsky/Corbis-Stock Market.

SRAonline.com

 **SRA**

Send all inquiries to this address:
SRA/McGraw-Hill
8787 Orion Place
Columbus, OH 43240

ISBN: 978-0-07-618619-8
MHID: 0-07-618619-9

5 6 7 8 9 QDB 15 14 13 12 11

Contents

ORANGE BOOK PRETESTS
Readiness Check

Add or subtract.

	a	*b*	*c*	*d*	*e*
1.	63 + 5	47 +32	60 +58	85 +69	563 +97
2.	5632 +798	8067 +5978	5027 893 +1605	6753 8965 +4786	$56.85 +28.97
3.	14 − 8	47 −15	73 − 9	64 −27	270 −56
4.	716 −409	800 −173	5027 −688	9132 −1698	13600 −7516

Multiply.

	a	*b*	*c*	*d*	*e*
5.	7 ×9	60 ×8	141 × 2	312 × 5	6093 × 8
6.	42 ×56	509 ×78	264 ×123	706 ×148	3576 × 23

ORANGE BOOK PRETESTS
Readiness Check (continued)

Divide.

	a	*b*	*c*	*d*
7.	4)‾2‾8‾	6)‾5‾4‾	7)‾5‾6‾	9)‾0‾
8.	8)‾3‾2‾	5)‾4‾5‾	9)‾7‾2‾	7)‾4‾2‾
9.	6)‾4‾8‾	9)‾6‾3‾	8)‾6‾4‾	7)‾4‾9‾

Complete the following.

	a	*b*
10.	1 quart = _____ pints	1 yard = _____ feet
11.	1 gallon = _____ quarts	1 hour = _____ minutes
12.	1 foot = _____ inches	1 day = _____ hours

Write the numerals that tell the time.

a   *b*

13.

_____ : _____   _____ : _____

Solve.

14. Juan has saved $47.25 to buy a bicycle. The bicycle costs $85.50. How much more does he need to buy the bicycle?

He needs _____ more.

ORANGE BOOK PRETESTS
Addition Facts (Pretest 1)

	a	*b*	*c*	*d*	*e*	*f*	*g*	*h*
1.	5 +7	0 +1	7 +2	4 +1	1 +6	3 +0	9 +5	8 +4
2.	5 +4	1 +5	8 +5	2 +9	4 +6	2 +0	6 +9	6 +3
3.	9 +4	1 +1	4 +2	6 +4	3 +9	9 +6	5 +3	0 +4
4.	2 +2	5 +6	3 +3	0 +0	7 +1	8 +3	6 +2	8 +0
5.	7 +3	0 +6	5 +2	7 +9	8 +2	9 +3	7 +7	2 +8
6.	6 +7	8 +6	4 +4	3 +4	2 +3	7 +0	5 +0	1 +3
7.	6 +5	9 +9	8 +7	5 +8	0 +9	7 +8	4 +9	3 +5
8.	5 +9	2 +5	7 +4	3 +6	4 +5	3 +8	7 +6	4 +7
9.	1 +2	5 +5	8 +8	9 +2	4 +8	9 +1	8 +9	2 +7
10.	3 +7	6 +8	2 +6	9 +8	6 +6	1 +8	9 +7	7 +5

ORANGE BOOK PRETESTS
Addition Facts (Pretest 2)

	a	*b*	*c*	*d*	*e*	*f*	*g*	*h*
1.	2 +1	8 +5	6 +5	4 +6	1 +0	7 +2	5 +5	3 +3
2.	8 +4	0 +0	7 +3	3 +5	9 +4	5 +1	2 +9	6 +2
3.	5 +6	2 +2	4 +7	0 +2	9 +0	8 +3	8 +7	3 +6
4.	3 +1	8 +2	5 +2	3 +2	6 +1	1 +1	6 +6	9 +9
5.	9 +3	1 +4	7 +4	5 +7	4 +8	0 +3	8 +1	2 +8
6.	0 +8	2 +4	6 +7	5 +3	9 +8	7 +9	3 +7	6 +0
7.	7 +5	9 +6	1 +7	2 +5	8 +6	5 +8	4 +9	0 +5
8.	4 +4	6 +8	6 +3	3 +8	8 +9	7 +8	2 +7	4 +5
9.	3 +9	9 +7	0 +7	5 +9	1 +9	6 +9	4 +3	7 +7
10.	5 +4	4 +0	8 +8	7 +6	9 +2	2 +6	9 +5	6 +4

ORANGE BOOK PRETESTS
Subtraction Facts (Pretest 1)

	a	*b*	*c*	*d*	*e*	*f*	*g*	*h*
1.	6 −2	1 1 −7	3 −3	1 3 −5	6 −1	1 1 −6	0 −0	1 3 −8
2.	6 −3	1 2 −6	5 −4	1 0 −6	7 −2	1 3 −9	4 −1	1 0 −7
3.	5 −2	1 0 −9	8 −2	1 2 −9	7 −3	1 1 −5	9 −6	1 7 −8
4.	8 −4	1 2 −5	9 −1	1 0 −3	9 −7	1 3 −4	1 0 −1	1 5 −9
5.	7 −5	1 1 −9	3 −1	1 4 −8	8 −3	1 0 −5	1 2 −3	1 0 −8
6.	2 −1	1 4 −5	6 −5	1 2 −8	7 −6	1 1 −3	1 0 −4	1 4 −7
7.	9 −9	1 2 −7	4 −0	1 3 −7	1 −1	1 6 −9	5 −5	1 5 −6
8.	8 −1	1 5 −8	9 −4	1 3 −6	7 −0	1 1 −2	9 −3	1 6 −7
9.	9 −2	1 4 −6	6 −0	1 7 −9	8 −8	1 2 −4	1 0 −2	1 1 −8
10.	9 −0	1 8 −9	4 −2	1 5 −7	2 −0	1 6 −8	1 1 −4	1 4 −9

ORANGE BOOK PRETESTS
Subtraction Facts (Pretest 2)

	a	*b*	*c*	*d*	*e*	*f*	*g*	*h*
1.	12 −3	9 −5	15 −6	7 −3	10 −4	9 −2	5 −5	12 −6
2.	11 −2	8 −4	10 −5	7 −7	16 −8	8 −0	10 −3	10 −9
3.	17 −8	6 −3	16 −9	5 −3	10 −7	8 −2	9 −4	14 −5
4.	11 −3	9 −8	11 −7	6 −6	13 −9	2 −1	7 −2	13 −6
5.	15 −9	2 −2	11 −5	0 −0	10 −8	6 −5	13 −4	12 −7
6.	10 −2	7 −4	12 −9	4 −4	10 −1	8 −7	3 −0	15 −8
7.	11 −4	1 −0	14 −8	4 −3	17 −9	9 −3	13 −7	11 −8
8.	14 −7	8 −6	11 −6	5 −0	12 −5	6 −4	5 −1	16 −7
9.	11 −9	7 −1	13 −8	3 −2	10 −6	9 −6	18 −9	13 −5
10.	12 −4	9 −9	14 −6	8 −5	15 −7	4 −2	12 −8	14 −9

ORANGE BOOK PRETESTS
Multiplication Facts (Pretest 1)

	a	*b*	*c*	*d*	*e*	*f*	*g*	*h*
1.	8 ×2	2 ×3	7 ×3	4 ×5	3 ×1	5 ×2	2 ×9	7 ×0
2.	5 ×1	8 ×0	1 ×1	4 ×6	2 ×0	9 ×1	6 ×3	0 ×9
3.	4 ×7	0 ×0	7 ×1	3 ×3	8 ×3	5 ×3	4 ×0	9 ×2
4.	2 ×4	3 ×9	6 ×2	9 ×7	0 ×1	8 ×4	1 ×7	2 ×2
5.	9 ×8	1 ×4	8 ×5	5 ×4	7 ×9	6 ×1	7 ×4	2 ×8
6.	5 ×5	8 ×6	4 ×8	7 ×5	6 ×4	3 ×5	6 ×9	9 ×6
7.	1 ×9	7 ×8	0 ×5	8 ×7	1 ×2	5 ×8	3 ×8	6 ×0
8.	5 ×9	2 ×7	8 ×8	9 ×9	9 ×5	6 ×5	8 ×9	5 ×7
9.	9 ×4	7 ×6	3 ×6	6 ×8	1 ×8	0 ×6	4 ×3	9 ×3
10.	3 ×7	0 ×7	4 ×4	5 ×6	4 ×9	2 ×6	7 ×7	6 ×6

ORANGE BOOK PRETESTS
Multiplication Facts (Pretest 2)

	a	*b*	*c*	*d*	*e*	*f*	*g*	*h*
1.	3 ×8	2 ×2	7 ×3	3 ×2	0 ×6	6 ×3	2 ×7	4 ×1
2.	3 ×3	0 ×0	6 ×2	2 ×1	5 ×2	3 ×0	7 ×4	1 ×0
3.	5 ×4	7 ×2	5 ×0	4 ×2	3 ×7	8 ×1	0 ×4	5 ×5
4.	3 ×9	2 ×8	1 ×1	5 ×3	7 ×5	4 ×9	8 ×9	1 ×9
5.	9 ×4	5 ×6	8 ×5	4 ×8	0 ×3	8 ×6	6 ×4	9 ×9
6.	1 ×3	9 ×2	2 ×9	7 ×6	9 ×8	5 ×7	4 ×3	0 ×8
7.	7 ×9	4 ×7	8 ×4	5 ×8	4 ×4	7 ×1	9 ×0	3 ×6
8.	6 ×8	3 ×4	0 ×2	9 ×3	1 ×5	7 ×7	6 ×5	8 ×2
9.	8 ×8	1 ×6	4 ×5	6 ×6	5 ×9	9 ×5	6 ×9	7 ×0
10.	4 ×6	6 ×7	9 ×7	7 ×8	8 ×3	3 ×5	1 ×8	9 ×6

ORANGE BOOK PRETESTS
Division Facts (Pretest 1)

	a	*b*	*c*	*d*	*e*	*f*	*g*
1.	5)̄1 0	7)̄7	6)̄3 0	8)̄2 4	4)̄1 2	9)̄9	5)̄4 0
2.	9)̄4 5	4)̄8	5)̄1 5	7)̄2 1	6)̄1 2	2)̄1 8	8)̄3 2
3.	2)̄4	8)̄1 6	4)̄2 8	9)̄5 4	3)̄1 8	7)̄5 6	3)̄0
4.	5)̄5	1)̄3	9)̄0	7)̄6 3	5)̄2 0	6)̄3 6	4)̄3 2
5.	9)̄3 6	4)̄2 4	3)̄6	8)̄4 0	2)̄6	3)̄1 2	9)̄6 3
6.	8)̄8	1)̄5	6)̄4 2	1)̄6	3)̄1 5	8)̄7 2	1)̄7
7.	1)̄2	6)̄1 8	2)̄1 6	4)̄3 6	2)̄8	1)̄9	7)̄0
8.	9)̄1 8	1)̄1	7)̄4 9	8)̄4 8	4)̄2 0	9)̄2 7	5)̄2 5
9.	2)̄1 0	6)̄4 8	3)̄9	9)̄7 2	2)̄0	7)̄2 8	3)̄3
10.	5)̄3 5	7)̄4 2	2)̄1 4	4)̄0	6)̄5 4	7)̄1 4	3)̄2 4
11.	1)̄8	6)̄0	3)̄2 7	5)̄4 5	3)̄2 1	2)̄2	8)̄5 6
12.	2)̄1 2	9)̄8 1	8)̄6 4	5)̄3 0	7)̄3 5	4)̄4	6)̄2 4

ORANGE BOOK PRETESTS
Division Facts (Pretest 2)

	a	b	c	d	e	f	g
1.	1)8	9)9	3)15	7)63	2)18	5)20	8)16
2.	4)28	5)25	6)6	2)0	9)18	3)18	6)48
3.	2)16	6)54	7)56	5)15	5)45	1)9	8)8
4.	6)0	2)2	6)18	3)0	9)27	4)0	4)36
5.	1)7	3)21	8)24	7)14	2)4	8)48	1)2
6.	8)56	7)49	4)32	3)12	6)24	2)6	9)36
7.	3)3	9)81	1)3	8)0	5)10	1)0	7)21
8.	7)0	5)30	4)24	2)8	3)24	8)40	4)16
9.	2)10	7)42	9)72	6)30	7)28	6)42	8)72
10.	4)20	1)5	3)27	3)9	5)35	2)12	9)45
11.	2)14	7)7	8)32	9)63	6)12	5)0	4)12
12.	1)4	4)4	8)64	7)35	6)36	9)54	5)40

NAME _____

ORANGE BOOK PRETESTS
Mixed Facts Pretest

Add, subtract, multiply, or divide. Watch the signs.

	a	*b*	*c*	*d*	*e*	*f*	*g*
1.	9)63	6 ×3	8 +8	2 +9	11 −3	16 −9	7 ×8
2.	7)35	10 −5	18 −9	7 +7	8 ×8	6 +6	3 ×9
3.	9)81	4 ×8	9 ×6	8 +5	5 +6	15 −6	13 −9
4.	5)45	7 ×7	9 +5	17 −8	5 ×6	8 +9	17 −9
5.	8)64	6 ×6	9 ×0	14 −7	4 +8	16 −8	7 +5
6.	8)40	15 −8	5 ×5	4 ×7	12 −5	7 +8	5 +4
7.	6)42	7 +6	8 +3	14 −9	9 ×9	13 −6	7 ×9
8.	7)56	3 ×4	9 +7	11 −7	6 ×4	9 +9	10 −6

Orange Book

ORANGE BOOK PRETESTS
Mixed Facts Pretest

11

Add, subtract, multiply, or divide.　Watch the signs.

	a	*b*	*c*	*d*
9.	34 +5	76 −4	32 × 3	9)54
10.	50 −30	16 ×7	54 +35	7)56
11.	267 ×4	8)97	80 −6	358 +49
12.	6)258	675 +894	241 −76	609 × 8
13.	6703 +472	302 −76	450 ×9	8)369
14.	$45.16 −21.78	$0.43 ×8	$37.50 +26.75	7)902

PROBLEM SOLVING STRATEGIES
Choose the Operation

Bert's Bed and Breakfast has about 35 guests every week. On average, how many guests does Bert have each day?

There are ____7____ days in a week.

__Divide__ to determine how many guests Bert has each day.

On average, Bert has ____5____ guests each day.

Divide 35 into 7 equal groups.

$$\begin{array}{r} 5 \\ 7\overline{)35} \end{array}$$

On average, each day of the week Bert will have 5 guests.

Solve each problem.

SHOW YOUR WORK

1. Nancy served 23 customers on Monday, 23 customers on Tuesday, 18 on Thursday, and 34 on Saturday. How many customers did Nancy serve in all?

   Do you add or subtract? _____

   Nancy served _____ customers.

2. Britanny bowls four times each month. How many times does she bowl in six months?

   Do you multiply or divide? _____

   Britanny bowls _____ times in six months.

3. To earn money, Brian shovels snow and pet sits. He earns $10 an hour shoveling snow and $4 an hour pet sitting. How much more an hour does Brian earn shoveling snow than he earns pet sitting?

   Do you add or subtract? _____

   Brian earns _____ more an hour shoveling snow.

PROBLEM SOLVING STRATEGIES
Draw a Picture

Tanya rode her bicycle 2 miles from her home to the store. After she went to the store, she rode 5 miles to the park and then 4 miles back home. How many miles did she ride in all?

2 miles	7 miles
+ 5 miles	+ 4 miles
7 miles	11 miles

Tanya rode ____11____ miles in all.

Draw a picture to help you find the answer.

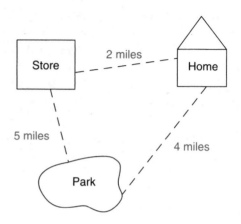

Solve each problem.

$\boxed{\text{SHOW YOUR WORK}}$

1. Marianne is making a beaded necklace. She wants the repeating pattern to be two striped beads, one red bead, then one black bead. If she has six striped beads and three red beads, how many black beads will she need to complete the necklace?

   She needs _____ black beads.

   There will be _____ beads in the necklace.

---

2. Larry planted five tulip bulbs in a circle. He planted two marigolds in between each of the tulip bulbs. How many marigolds did he plant?

   Larry planted _____ marigolds.

   How many flowers in all? _____

PROBLEM SOLVING STRATEGIES
Look for a Pattern

The pet store had 21 gerbils for sale on Monday, 18 on Tuesday, 15 on Wednesday, and 12 on Thursday. If this pattern continues, how many gerbils will the pet store have for sale on Friday?

There were ____3____ fewer gerbils for sale on Tuesday than on Monday.

There were ____3____ fewer gerbils for sale on Wednesday than on Tuesday.

There were ____3____ fewer gerbils for sale on Thursday than on Wednesday.

The pet store should have ____9____ gerbils for sale on Friday.

Look for a pattern.

Day:	M	T	W	Th	Fr
Number of gerbils:	21	18	15	12	9

The pattern is −3.

$12 - 3 = 9$

Solve each problem.

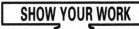

SHOW YOUR WORK

1. Kelly began with 48 tokens. She gave 24 to Kuri. She then gave 12 to Tani. If this pattern continues, how many will Kelly give away next?

    48 ÷ _____ = 24

    24 ÷ _____ = 12

    12 ÷ _____ = _____

    Kelly will give _____ tokens away next.

---

2. The furnace at Lucy's school broke. In the first hour, the temperature dropped 1 degree. In the second hour, the temperature dropped 3 degrees. It dropped 6 more degrees during the third hour and 10 more degrees during the fourth hour. If this pattern continues, how many more degrees will the temperature drop during the fifth hour?

    The temperature will drop another _____ degrees during the fifth hour.

PROBLEM SOLVING STRATEGIES
Guess and Check

There are 24 fourth graders on the playground. There are 8 more boys than girls. How many fourth-grade boys are on the playground? How many fourth-grade girls are on the playground?

There are ____16____ fourth-grade boys on the playground.

There are ____8____ fourth-grade girls on the playground.

Guess possible numbers of boys and girls. Check to see if the sum of the numbers equals 24.

Guess: 12 girls and 20 boys
Sum: 12 + 20 = 32   Incorrect.

Guess: 8 girls and 16 boys
Sum: 8 + 16 = 24   Correct.

Solve each problem.

SHOW YOUR WORK

1. A movie ticket and popcorn cost Andrew $8.50. The popcorn costs $1.50 more than the ticket. How much did the ticket cost? How much did the popcorn cost?

   The ticket cost _____ .

   The popcorn cost _____ .

2. The distance around Regina's rectangular desk is 66 inches. The length of the desk is 3 inches more than the width. What are the length and width of Regina's desk?

   Regina's desk is _____ inches long.

   Regina's desk is _____ inches wide.

3. The sum of two numbers is 12. Their difference is 6. What are the two numbers?

   The two numbers are _____ and _____ .

PROBLEM SOLVING STRATEGIES
Identify Missing Information

During the first week of May, Sabrina and Jared collected newspapers for a paper drive. They collected 300 pounds on Monday, 150 pounds on Tuesday, and 238 pounds on Wednesday. How many more pounds did they need to collect on Thursday and Friday to meet their goal?

Not enough information

Missing information: the number of pounds in their goal

Add to find the total number of pounds collected on Monday, Tuesday, and Wednesday.

$$\begin{array}{r} 300 \\ 150 \\ +\ 238 \\ \hline 688 \text{ pounds collected} \end{array}$$

$$\begin{array}{r} \text{goal} \\ -\ 688 \text{ pounds collected} \\ \hline \text{pounds needed to collect} \end{array}$$

Information on the number of pounds that was their goal is missing.

Solve each problem.

SHOW YOUR WORK

1. Lee gave her cat, Tinker, 11 treats. She gave her other cat, Bell, 8 treats, and then she gave her aunt's cat the rest of the treats. How many treats did Lee have at first?

_____

Missing information: _____

_____

---

2. Four friends are in line to tour the White House. Cassie is in front of Dawan. Dawan is between Linda and Katie. Katie is last. In what order are the friends?

_____

Missing information: _____

_____

PROBLEM SOLVING STRATEGIES
Make a Table

Keisha's class sold magazines as a fund-raiser. For every $50 worth of sales brought in, the class keeps $20. Keisha's class brought in $300 worth of magazine sales. How much money does the class get to keep?

Keisha's class keeps _____$40_____ when they bring in $100.

Keisha's class keeps _____$60_____ when they bring in $150.

Keisha's class keeps _____$120_____ when they bring in $300.

Make a table.

Sales	Class Profit
$ 50	$ 20
$100	$ 40
$150	$ 60
$200	$ 80
$250	$100
$300	$120

Solve each problem.

| SHOW YOUR WORK |

1. To every soccer game, Mrs. Romer brings orange slices for her daughter's team. For every 4 soccer players, she needs three oranges. How many oranges does she need if the team has 16 players?

   Mrs. Romer will need _____ oranges for 8 soccer players.

   Mrs. Romer will need _____ oranges for 16 soccer players.

---

2. In every five packs of baseball cards, Jerome finds 3 rookie cards. How many packs of baseball cards will he need to open to find 18 rookie cards?

   Jerome will need to open _____ packs of cards to find 6 rookie cards.

   Jerome will need to open _____ packs of cards to find 18 rookie cards.

PROBLEM SOLVING STRATEGIES

Work a Simpler Problem

There are seven people meeting for the first time at the airport. Each person must shake hands with each of the other people. How many handshakes will there be in all?

Two people shake hands a total of ____1____ time.

Three people shake hands a total of ____3____ times.

Seven people shake hands a total of ____21____ times.

Solve the problem for fewer people. Start with two people.

Number of People	Number of Shakes
2	1
3	3
4	6
5	10
6	15
7	21

The pattern: +2, +3, +4

Solve each problem.

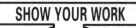

1. Mrs. Ling must buy the supplies for the school play. She needs 80 ribbons that cost $2 each for the girls, and 90 collars that cost $3 each for the boys. How much will she spend on supplies?

   _____ for the ribbons

   _____ for the collars

   Total cost: _____

---

2. Donna hired an electrician twice. The first time he worked four hours, and the second time he worked five hours. He charges $20 an hour. How much did she pay him in all?

   total hours worked _____

   Donna paid him _____ in all.

PROBLEM SOLVING STRATEGIES
Work Backward

Phyllis first cut off 18 centimeters from the length of a strip of paper. A few minutes later, she cut off another 5 centimeters. The paper is now 125 centimeters long. How long was the paper in the beginning?

Phyllis started with ____148____ centimeters of paper.

Work backward from the amount Phyllis had left to find the amount she had in the beginning.

Ending length:	125 cm
Cut off:	+ 5 cm
	130 cm
Cut off:	+ 18 cm
Beginning length:	148 cm

Solve each problem.

SHOW YOUR WORK

1. Leon ate lunch at his favorite cafeteria. He paid $0.99 for milk, $3.55 for a sandwich and fries, and $1.25 for a dish of fruit. He received $0.25 in change. How much money did he give the cashier?

   Leon gave the cashier _____ .

2. Peter and Motega played a number game. Peter picked a number and subtracted 2. Then he added 7. Next, Motega added 8. The result was 25. What number did Peter choose?

   Peter chose the number _____ .

3. Vivian loves to take pictures of animals. On a recent trip to the zoo, she took three pictures of the giraffes, seven pictures of the monkeys, and four pictures at the lion's den. She still had ten blank shots left. How many blank shots were on the roll of film when she came to the zoo?

   Vivian had _____ blank shots on the roll of film when she went to the zoo.

CHAPTER 1 PRETEST
Addition and Subtraction (1- and 2-digit; no renaming)

Add.

	a	*b*	*c*	*d*	*e*	*f*
1.	45 +4	73 +5	64 +5	22 +7	42 +5	74 +4
2.	4 +43	3 +54	6 +43	3 +34	5 +24	2 +46
3.	70 +10	60 +20	40 +50	40 +30	70 +20	30 +40
4.	54 +21	23 +41	11 +16	22 +23	45 +21	53 +32
5.	68 +30	11 +25	34 +30	12 +80	12 +71	45 +40

Subtract.

	a	*b*	*c*	*d*	*e*	*f*
6.	28 −7	57 −6	46 −4	29 −8	47 −5	78 −6
7.	50 −10	40 −20	60 −40	70 −30	90 −60	80 −30
8.	89 −46	97 −94	89 −55	78 −23	98 −48	67 −23
9.	78 −70	98 −32	99 −21	67 −61	89 −10	66 −51

CHAPTER 1 PRETEST   Problem Solving

Solve each problem.

1. Heath took three pictures of red flowers and two pictures of yellow flowers. How many pictures of flowers did he take in all?

He took _____ pictures of red flowers.

He took _____ pictures of yellow flowers.

He took _____ pictures of flowers in all.

2. Heath also took nine pictures of people swimming. Only five of the pictures had seagulls in them. How many pictures did not have seagulls in them?

There were _____ pictures of people swimming.

_____ of the pictures had seagulls in them.

_____ of the pictures did not have seagulls in them.

3. During the day Heath used four rolls of black-and-white film and three rolls of color film. How many rolls of film did he use in all?

He used _____ rolls of black-and-white film.

He used _____ rolls of color film.

He used _____ rolls of film in all.

1.

2.

3.

Lesson 1   Addition Facts

4 → Find the **4**-row.

+ 5 → Find the **5**-column.

9 ← The sum is named where the 4-row and 5-column meet.

Use the table to add.

```
  7
+ 8
```

+	0	1	2	3	4	5	6	7	8	9
0	0	1	2	3	4	5	6	7	8	9
1	1	2	3	4	5	6	7	8	9	10
2	2	3	4	5	6	7	8	9	10	11
3	3	4	5	6	7	8	9	10	11	12
4	4	5	6	7	8	9	10	11	12	13
5	5	6	7	8	9	10	11	12	13	14
6	6	7	8	9	10	11	12	13	14	15
7	7	8	9	10	11	12	13	14	15	16
8	8	9	10	11	12	13	14	15	16	17
9	9	10	11	12	13	14	15	16	17	18

Add.

	a	b	c	d	e	f	g	h
1.	5 +3	2 +5	5 +4	6 +3	3 +4	3 +5	2 +7	3 +7
2.	6 +2	3 +6	3 +2	2 +4	4 +3	2 +6	7 +2	4 +4
3.	2 +3	0 +7	3 +1	9 +0	1 +8	0 +5	4 +2	6 +1
4.	7 +9	6 +7	9 +4	8 +3	4 +9	7 +3	8 +5	8 +9
5.	7 +7	9 +5	4 +7	6 +5	2 +8	7 +5	4 +8	7 +6
6.	9 +6	5 +8	5 +9	6 +6	9 +8	7 +4	3 +9	2 +9
7.	4 +6	1 +9	5 +7	9 +3	3 +8	8 +4	9 +7	9 +9

Orange Book

Lesson 1
Addition Facts

23

Lesson 2 Subtraction Facts

9 → Find the 9 in

− 6 { the 6 -column.

3 ← The difference is named in the ▓ at the end of this row.

Use the table to subtract.

$$\begin{array}{r} 12 \\ -\ 4 \\ \hline \end{array}$$

−	0	1	2	3	4	5	6	7	8	9
0	0	1	2	3	4	5	6	7	8	9
1	1	2	3	4	5	6	7	8	9	10
2	2	3	4	5	6	7	8	9	10	11
3	3	4	5	6	7	8	9	10	11	12
4	4	5	6	7	8	9	10	11	12	13
5	5	6	7	8	9	10	11	12	13	14
6	6	7	8	9	10	11	12	13	14	15
7	7	8	9	10	11	12	13	14	15	16
8	8	9	10	11	12	13	14	15	16	17
9	9	10	11	12	13	14	15	16	17	18

Subtract.

	a	b	c	d	e	f	g	h
1.	7 −6	9 −7	7 −4	3 −2	8 −0	9 −5	8 −7	7 −1
2.	5 −3	6 −5	7 −0	8 −2	8 −4	9 −9	8 −5	5 −2
3.	13 −6	10 −2	14 −5	12 −5	11 −6	10 −9	18 −9	11 −9
4.	15 −9	11 −8	16 −8	12 −6	14 −6	11 −3	15 −8	12 −8
5.	16 −9	13 −4	15 −7	10 −4	12 −9	13 −5	11 −7	10 −6
6.	14 −9	11 −2	10 −3	14 −8	16 −7	14 −7	17 −8	11 −4
7.	13 −9	13 −7	10 −8	15 −6	13 −8	12 −7	17 −9	11 −5

Lesson 3 Addition and Subtraction (1- and 2-digit)

	Add the ones.	Add the tens.		Subtract the ones.	Subtract the tens.
36 +2	36 +2 — 8	36 +2 — 38	57 −4 —	57 −4 — 3	57 −4 — 53

Add.

	a	*b*	*c*	*d*	*e*	*f*
1.	2 0 +7	3 1 +1	4 2 +1	1 8 +1	2 1 +6	5 4 +4
2.	2 2 +1	4 2 +2	6 3 +3	7 5 +4	6 4 +5	9 6 +3
3.	3 +6 6	2 +1 6	2 +1 7	1 +2 8	9 +3 0	7 +3 2
4.	7 +4 1	8 +6 1	2 +3 5	6 +4 1	8 +4 0	5 +7 1

Subtract.

5.	1 9 −8	3 8 −7	5 9 −5	2 7 −6	4 5 −3	9 6 −5
6.	9 7 −5	4 9 −6	1 8 −5	6 9 −7	3 8 −6	5 9 −9
7.	8 9 −4	2 8 −4	4 6 −4	7 7 −4	8 9 −3	6 5 −2
8.	7 9 −1	6 8 −2	3 9 −2	8 6 −3	5 7 −3	7 8 −3

Lesson 3 Problem Solving

Solve each problem.

1. Regina worked 27 hours. Louis worked 6. How many more hours did Regina work than Louis?

   Regina worked _____ hours.

   Louis worked _____ hours.

   Regina worked _____ more hours.

2. The Cubs got 8 hits in the first game. They got 11 hits in the second game. How many hits did they get in both games?

   They got _____ hits in the first game.

   They got _____ hits in the second game.

   They got _____ hits in both games.

3. Five of Mr. Spangler's workers are absent today. Twenty-three of them are at work today. How many workers does he have in all?

   _____ workers are present.

   _____ workers are absent.

   He has _____ workers.

4. Melody has 12 CDs. Paul has 6 CDs. How many CDs do they have in all?

   Melody has _____ CDs.

   Paul has _____ CDs.

   They have _____ CDs in all.

5. There are 19 women and 7 men in Zachary's apartment building. How many more women than men live in that apartment building?

   _____ more women live in the building.

1.	
2.	
3.	
4.	5.

Lesson 4   Addition (2-digit)

$$\begin{array}{r} 6 \\ +2 \\ \hline 8 \end{array} \qquad \begin{array}{r} 60 \\ +20 \\ \hline 80 \end{array}$$

If 6 + 2 = 8,
then 60 + 20 = 80.

	Add the ones.	Add the tens.
$\begin{array}{r} 36 \\ +21 \\ \hline \end{array}$	$\begin{array}{r} 36 \\ +21 \\ \hline 7 \end{array}$	$\begin{array}{r} 36 \\ +21 \\ \hline 57 \end{array}$

Add.

	a	b	c	d	e	f
1.	$\begin{array}{r} 5 \\ +1 \\ \hline \end{array}$	$\begin{array}{r} 50 \\ +10 \\ \hline \end{array}$	$\begin{array}{r} 4 \\ +3 \\ \hline \end{array}$	$\begin{array}{r} 40 \\ +30 \\ \hline \end{array}$	$\begin{array}{r} 7 \\ +2 \\ \hline \end{array}$	$\begin{array}{r} 70 \\ +20 \\ \hline \end{array}$
2.	$\begin{array}{r} 3 \\ +6 \\ \hline \end{array}$	$\begin{array}{r} 30 \\ +60 \\ \hline \end{array}$	$\begin{array}{r} 5 \\ +2 \\ \hline \end{array}$	$\begin{array}{r} 50 \\ +20 \\ \hline \end{array}$	$\begin{array}{r} 1 \\ +8 \\ \hline \end{array}$	$\begin{array}{r} 10 \\ +80 \\ \hline \end{array}$
3.	$\begin{array}{r} 60 \\ +30 \\ \hline \end{array}$	$\begin{array}{r} 40 \\ +40 \\ \hline \end{array}$	$\begin{array}{r} 70 \\ +10 \\ \hline \end{array}$	$\begin{array}{r} 20 \\ +60 \\ \hline \end{array}$	$\begin{array}{r} 50 \\ +30 \\ \hline \end{array}$	$\begin{array}{r} 40 \\ +20 \\ \hline \end{array}$
4.	$\begin{array}{r} 17 \\ +41 \\ \hline \end{array}$	$\begin{array}{r} 28 \\ +61 \\ \hline \end{array}$	$\begin{array}{r} 32 \\ +35 \\ \hline \end{array}$	$\begin{array}{r} 46 \\ +41 \\ \hline \end{array}$	$\begin{array}{r} 28 \\ +40 \\ \hline \end{array}$	$\begin{array}{r} 15 \\ +71 \\ \hline \end{array}$
5.	$\begin{array}{r} 22 \\ +14 \\ \hline \end{array}$	$\begin{array}{r} 33 \\ +64 \\ \hline \end{array}$	$\begin{array}{r} 44 \\ +33 \\ \hline \end{array}$	$\begin{array}{r} 85 \\ +13 \\ \hline \end{array}$	$\begin{array}{r} 71 \\ +24 \\ \hline \end{array}$	$\begin{array}{r} 71 \\ +13 \\ \hline \end{array}$
6.	$\begin{array}{r} 21 \\ +73 \\ \hline \end{array}$	$\begin{array}{r} 24 \\ +62 \\ \hline \end{array}$	$\begin{array}{r} 25 \\ +72 \\ \hline \end{array}$	$\begin{array}{r} 16 \\ +82 \\ \hline \end{array}$	$\begin{array}{r} 42 \\ +43 \\ \hline \end{array}$	$\begin{array}{r} 53 \\ +42 \\ \hline \end{array}$
7.	$\begin{array}{r} 63 \\ +36 \\ \hline \end{array}$	$\begin{array}{r} 58 \\ +31 \\ \hline \end{array}$	$\begin{array}{r} 64 \\ +24 \\ \hline \end{array}$	$\begin{array}{r} 47 \\ +42 \\ \hline \end{array}$	$\begin{array}{r} 76 \\ +13 \\ \hline \end{array}$	$\begin{array}{r} 35 \\ +63 \\ \hline \end{array}$

Lesson 4 Problem Solving

Solve each problem.

1. Eileen's family used 14 liters of milk this week. Last week they used 15 liters. How many liters of milk did they use in the two weeks?

   _____ liters were used this week.

   _____ liters were used last week.

   _____ liters were used in the two weeks.

2. Tom drove 42 kilometers today. He drove 46 kilometers yesterday. How many kilometers did he drive during the two days?

   _____ kilometers were driven today.

   _____ kilometers were driven yesterday.

   _____ kilometers were driven both days.

3. Our family has two dogs. Pepper weighs 31 pounds. Salt weighs 28 pounds. What is the combined weight of the two dogs?

   Pepper weighs _____ pounds.

   Salt weighs _____ pounds.

   The combined weight is _____ pounds.

4. Leslie scored 43 points. Michael scored 25 points. How many points did they score in all?

   Leslie scored _____ points.

   Michael scored _____ points.

   They scored _____ points in all.

5. Mr. Cook was 25 years old when Mary was born. How old will he be when Mary has her 13ᵗʰ birthday?

   Mr. Cook will be _____ years old.

1.

2.

3.

4.

5.

Lesson 5 Subtraction (2-digit)

$$\begin{array}{r} 5 \\ -3 \\ \hline 2 \end{array} \qquad \begin{array}{r} 50 \\ -30 \\ \hline 20 \end{array}$$

If 5 − 3 = 2,
then 50 − 30 = 20.

	Subtract the ones.	Subtract the tens.
$\begin{array}{r} 67 \\ -53 \\ \hline \end{array}$	$\begin{array}{r} 67 \\ -53 \\ \hline 4 \end{array}$	$\begin{array}{r} 67 \\ -53 \\ \hline 14 \end{array}$

Subtract.

	a	b	c	d	e	f
1.	$\begin{array}{r} 4 \\ -2 \\ \hline \end{array}$	$\begin{array}{r} 40 \\ -20 \\ \hline \end{array}$	$\begin{array}{r} 7 \\ -4 \\ \hline \end{array}$	$\begin{array}{r} 70 \\ -40 \\ \hline \end{array}$	$\begin{array}{r} 9 \\ -6 \\ \hline \end{array}$	$\begin{array}{r} 90 \\ -60 \\ \hline \end{array}$
2.	$\begin{array}{r} 8 \\ -3 \\ \hline \end{array}$	$\begin{array}{r} 80 \\ -30 \\ \hline \end{array}$	$\begin{array}{r} 7 \\ -5 \\ \hline \end{array}$	$\begin{array}{r} 70 \\ -50 \\ \hline \end{array}$	$\begin{array}{r} 6 \\ -2 \\ \hline \end{array}$	$\begin{array}{r} 60 \\ -20 \\ \hline \end{array}$
3.	$\begin{array}{r} 40 \\ -30 \\ \hline \end{array}$	$\begin{array}{r} 90 \\ -70 \\ \hline \end{array}$	$\begin{array}{r} 80 \\ -40 \\ \hline \end{array}$	$\begin{array}{r} 70 \\ -20 \\ \hline \end{array}$	$\begin{array}{r} 60 \\ -40 \\ \hline \end{array}$	$\begin{array}{r} 50 \\ -10 \\ \hline \end{array}$
4.	$\begin{array}{r} 98 \\ -87 \\ \hline \end{array}$	$\begin{array}{r} 76 \\ -66 \\ \hline \end{array}$	$\begin{array}{r} 89 \\ -61 \\ \hline \end{array}$	$\begin{array}{r} 57 \\ -45 \\ \hline \end{array}$	$\begin{array}{r} 49 \\ -39 \\ \hline \end{array}$	$\begin{array}{r} 65 \\ -53 \\ \hline \end{array}$
5.	$\begin{array}{r} 69 \\ -40 \\ \hline \end{array}$	$\begin{array}{r} 78 \\ -45 \\ \hline \end{array}$	$\begin{array}{r} 69 \\ -37 \\ \hline \end{array}$	$\begin{array}{r} 45 \\ -22 \\ \hline \end{array}$	$\begin{array}{r} 87 \\ -83 \\ \hline \end{array}$	$\begin{array}{r} 95 \\ -60 \\ \hline \end{array}$
6.	$\begin{array}{r} 53 \\ -51 \\ \hline \end{array}$	$\begin{array}{r} 94 \\ -51 \\ \hline \end{array}$	$\begin{array}{r} 84 \\ -44 \\ \hline \end{array}$	$\begin{array}{r} 98 \\ -43 \\ \hline \end{array}$	$\begin{array}{r} 27 \\ -12 \\ \hline \end{array}$	$\begin{array}{r} 66 \\ -20 \\ \hline \end{array}$
7.	$\begin{array}{r} 86 \\ -21 \\ \hline \end{array}$	$\begin{array}{r} 74 \\ -10 \\ \hline \end{array}$	$\begin{array}{r} 95 \\ -31 \\ \hline \end{array}$	$\begin{array}{r} 83 \\ -12 \\ \hline \end{array}$	$\begin{array}{r} 49 \\ -22 \\ \hline \end{array}$	$\begin{array}{r} 78 \\ -50 \\ \hline \end{array}$

Lesson 5 Problem Solving

Solve each problem.

1. Catherine weighs 39 kilograms. Her sister weighs 23 kilograms. How much more does Catherine weigh?

Catherine weighs _____ kilograms.

Her sister weighs _____ kilograms.

Catherine weighs _____ kilograms more.

2. Jessica can kick a football 23 yards. She can throw it 21 yards. How much farther can she kick the football?

She can kick the football _____ yards.

She can throw the football _____ yards.

She can kick the football _____ yards farther.

3. Marcos has 47 dollars. He plans to spend 25 dollars on presents. How much money will he have left?

Marcos has _____ dollars.

He plans to spend _____ dollars.

He will have _____ dollars left.

4. Travis is running in a 26-mile race. So far he has run 16 miles. How many more miles must he run?

The race is _____ miles long.

Travis has gone _____ miles.

He has _____ more miles to go.

5. Arlene read 24 pages on Monday. She read 37 pages on Tuesday. How many more pages did she read on Tuesday than on Monday?

Arlene read _____ more pages on Tuesday.

1.

2.

3.

4. 5.

CHAPTER 1 PRACTICE TEST
Addition and Subtraction (1- and 2-digit; no renaming)

Add.

	a	*b*	*c*	*d*	*e*	*f*
1.	20 +9	28 +1	51 +7	52 +7	42 +7	23 +6
2.	5 +31	2 +81	3 +31	3 +94	4 +31	7 +51
3.	40 +20	10 +60	20 +60	40 +20	30 +20	50 +30
4.	35 +60	45 +22	43 +40	45 +33	41 +55	35 +14
5.	62 +34	75 +13	73 +25	64 +23	43 +25	25 +74

Subtract.

	a	*b*	*c*	*d*	*e*
6.	68 −6	29 −1	47 −6	59 −2	78 −5
7.	70 −30	60 −50	90 −40	80 −30	60 −40
8.	88 −78	77 −34	98 −50	67 −41	98 −91
9.	96 −62	88 −23	76 −23	99 −78	87 −40

CHAPTER 2 PRETEST
Addition and Subtraction (2- and 3-digit; with renaming)

Add.

	a	*b*	*c*	*d*	*e*	*f*
1.	1 2 +5	6 4 +2	7 8 +5	3 7 +5	8 7 +9	8 5 +4
2.	2 8 +7	4 2 +9	6 9 +5	7 +6 3	8 +4 7	7 +4 9
3.	6 7 +2 5	3 4 +7 2	8 7 +2 9	3 6 +2 1 5	2 9 5 +4 2	3 8 8 +2 5
4.	5 2 2 4 +3 7	6 2 5 7 +9 3	4 8 6 3 +7 5	9 6 5 2 +4 8	4 6 6 3 +7 2	9 6 5 1 +3 9

Subtract.

	a	*b*	*c*	*d*	*e*	*f*
5.	2 6 −7	8 3 −9	6 7 −8	4 3 −7	9 4 −6	7 6 −9
6.	4 3 −2 8	6 4 −2 5	7 3 −4 7	8 3 −2 4	7 6 −2 9	8 3 −2 5
7.	1 2 9 −6 7	8 3 9 −9 7	4 2 5 −8 4	9 8 7 −9 5	6 5 4 −5 7	8 3 2 −5 1
8.	3 6 4 −7 6	5 2 4 −6 9	6 0 8 −5 9	8 3 2 −4 7	7 8 4 −8 5	4 7 7 −9 8
9.	5 8 4 −7 5	8 3 4 −1 5	6 7 9 −8 9	9 8 3 −9 1	2 0 1 −9 6	4 8 5 −9 8

Lesson 1 Addition (three or more numbers)

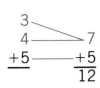

$$\begin{array}{r} 3 \\ 4 \\ +5 \\ \hline \end{array} \quad \begin{array}{r} 7 \\ +5 \\ \hline 12 \end{array}$$

$$\begin{array}{r} 3 \\ 1 \\ 2 \\ +5 \\ \hline \end{array} \quad \begin{array}{r} 4 \\ 2 \\ +5 \\ \hline \end{array} \quad \begin{array}{r} 6 \\ +5 \\ \hline 11 \end{array}$$

Add.

	a	b	c	d	e	f	g	h
1.	1 5 +7	1 8 +3	7 2 +5	4 5 +9	7 1 +9	2 2 +8	7 2 +8	6 3 +5
2.	3 2 +5	1 7 +5	6 2 +5	2 4 +5	3 3 +9	3 5 +7	4 4 +5	3 6 +4
3.	1 4 +8	5 3 +8	2 6 +8	4 1 +9	2 5 +4	5 4 +2	3 5 +3	6 3 +2
4.	2 3 3 +5	1 3 4 +8	6 2 1 +7	5 3 1 +4	4 4 1 +8	7 1 1 +9	6 2 1 +7	2 3 4 +9
5.	1 2 6 +9	1 3 5 +8	2 1 5 +7	2 2 5 +8	3 2 4 +7	3 1 2 +9	4 1 4 +6	2 3 2 +3
6.	3 5 2 +7	4 6 1 +9	3 5 5 +6	5 4 3 +7	7 1 2 +6	4 3 3 +8	8 3 1 +6	7 6 8 +5

Lesson 1   Problem Solving

Solve each problem.

1. A football team gained 2 yards on first down, 3 yards on second down, and 6 yards on third down. How many yards did they gain in the three downs?

   They gained _____ yards on first down.

   They gained _____ yards on second down.

   They gained _____ yards on third down.

   They gained _____ yards in three downs.

1.

2. Mr. Bernardo bought five apples, four oranges, and six pears. How many pieces of fruit did he buy?

   He bought _____ apples.

   He bought _____ oranges.

   He bought _____ pears.

   He bought _____ pieces of fruit.

2.

3. During a hike, Bethany counted three maple trees, four oak trees, two poplar trees, and seven aspen trees. How many trees did Bethany count?

   Bethany counted _____ trees.

3.

4. During a game, Amy made four baskets, Erin made two, Tina made three, and Kim made six. How many baskets did these girls make?

   The girls made _____ baskets.

4.

5. One piece of rope is 3 meters long. Another is 6 meters long. A third piece is 2 meters long. What is the combined length of the three pieces?

   The combined length is _____ meters.

5.

6. Carlotta was absent from school four days in March, six days in April, zero days in May, and three days in June. How many days was she absent during the four months?

   Carlotta was absent _____ days.

6.

Lesson 2 Addition (1- and 2-digit)

	Add the ones.	Add the tens.

```
                              1            1
   46                        46           46
  +27    6 + 7 = 13 or 10 + 3   +27          +27
                              3           73
```

Add.

	a	b	c	d	e	f
1.	17 +9	14 +8	32 +9	25 +6	31 +9	53 +8
2.	3 +28	7 +64	9 +56	5 +37	4 +38	8 +65
3.	32 +28	78 +15	65 +16	74 +18	29 +55	65 +27
4.	25 +46	27 +14	16 +25	17 +33	17 +26	28 +44
5.	38 +45	47 +39	63 +28	39 +52	44 +29	73 +18
6.	37 +36	38 +19	45 +29	58 +19	16 +27	19 +27
7.	35 +48	16 +75	32 +49	27 +38	29 +58	51 +29

Lesson 2 Problem Solving

Solve each problem.

1. The station has 16 whitewall tires. It has 9 black-wall tires. How many tires does the station have?

   There are _____ whitewall tires.

   There are _____ black-wall tires.

   There are _____ tires in all.

2. It takes 8 minutes to grease a car. It takes 15 minutes to change the oil. How long would it take to grease a car and change the oil?

   It would take _____ minutes to grease a car and change the oil.

3. This morning 26 people bought gasoline. This afternoon 37 people bought gasoline. How many people have bought gasoline today?

   _____ people have bought gasoline today.

4. Mrs. Verdugo bought 54 liters of gasoline on Tuesday. She bought 38 liters of gasoline on Friday. How many liters did she buy in all?

   She bought _____ liters of gasoline.

1.

2.

3.

4.

Lesson 3 Addition (2-digit)

Add the ones.   Add the tens.

$$\begin{array}{r} 48 \\ 57 \\ +29 \\ \hline \end{array} \qquad \begin{array}{r} \overset{2}{4}8 \\ 57 \\ +29 \\ \hline 4 \end{array} \qquad \begin{array}{r} \overset{2}{4}8 \\ 57 \\ +29 \\ \hline 134 \end{array}$$

$8 + 7 + 9 = 24$ $20 + 40 + 50 + 20 = 130$

$24 = 20 + 4$ $130 = 100 + 30$

Add.

	a	b	c	d	e	f
1.	7 3 +5 2	6 4 +9 3	6 5 +6 4	5 1 +7 8	8 2 +3 4	6 0 +5 7
2.	4 7 +8 1	5 6 +8 2	7 8 +4 1	8 4 +9 2	7 6 +8 3	8 6 +5 3
3.	2 8 +7 3	3 9 +9 2	8 7 +7 3	9 9 +5 1	7 9 +5 3	5 6 +7 5
4.	6 2 5 4 +7 2	7 2 8 3 +5 1	9 5 6 1 +2 2	9 5 8 2 +7 1	5 2 2 1 +6 6	3 2 2 3 +9 4
5.	6 2 7 5 +4 3	8 3 7 5 +6 4	9 6 8 7 +4 2	8 6 7 8 +5 4	6 7 3 2 +4 5	8 4 2 3 +5 7
6.	6 1 2 2 +1 7	6 3 5 7 +8 3	9 6 4 8 +7 6	2 7 8 3 +9 1	8 1 8 9 +9 5	7 5 3 6 +2 4

Lesson 3   Problem Solving

Solve each problem.

1. Mr. Ford has 61 hens and 54 roosters. How many
   chickens does he have?

   He has _____ hens.

   He has _____ roosters.

   He has _____ chickens.

2. Last week Tony read a 76-page book. This week he
   read an 83-page book. How many pages are in the
   two books?

   Last week he read a _____-page book.

   This week he read an _____-page book.

   There are _____ pages in the two books.

3. Jackie has 75 feet of kite string. Mike has 96 feet
   of kite string. How much kite string do they have?

   They have _____ feet of kite string.

4. A train went 57 kilometers the first hour. It
   went 65 kilometers the second hour. It went
   52 kilometers the third hour. How far did it go?

   The train went _____ kilometers.

5. A restaurant used 76 kilograms of potatoes,
   14 kilograms of carrots, and 68 kilograms of meat.
   What was the weight of the three items?

   The weight was _____ kilograms.

6. A cafeteria served 52 women, 47 men, and
   69 children. How many people did the cafeteria
   serve?

   _____ people were served.

1.

2.

3.

4.

5.

6.

Lesson 4   Subtraction (1-, 2-, and 3-digit)

To subtract the ones, rename 4 tens and 7 ones as "3 tens and 17 ones."

$$\begin{array}{r} 147 \\ -19 \\ \hline \end{array}$$

$$\begin{array}{r} \overset{3\ 17}{1\cancel{4}\cancel{7}} \\ -1\ 9 \\ \hline \end{array}$$

Subtract the ones.

$$\begin{array}{r} \overset{3\ 17}{1\cancel{4}\cancel{7}} \\ -1\ 9 \\ \hline 8 \end{array}$$

Subtract the tens.

$$\begin{array}{r} \overset{3\ 17}{1\cancel{4}\cancel{7}} \\ -1\ 9 \\ \hline 2\ 8 \end{array}$$

Subtract the hundreds.

$$\begin{array}{r} \overset{3\ 17}{1\cancel{4}\cancel{7}} \\ -1\ 9 \\ \hline 1\ 2\ 8 \end{array}$$

Subtract.

	a	b	c	d	e	f
1.	$\begin{array}{r} 26 \\ -8 \\ \hline \end{array}$	$\begin{array}{r} 63 \\ -7 \\ \hline \end{array}$	$\begin{array}{r} 72 \\ -9 \\ \hline \end{array}$	$\begin{array}{r} 42 \\ -5 \\ \hline \end{array}$	$\begin{array}{r} 83 \\ -6 \\ \hline \end{array}$	$\begin{array}{r} 75 \\ -7 \\ \hline \end{array}$
2.	$\begin{array}{r} 23 \\ -17 \\ \hline \end{array}$	$\begin{array}{r} 65 \\ -16 \\ \hline \end{array}$	$\begin{array}{r} 83 \\ -27 \\ \hline \end{array}$	$\begin{array}{r} 32 \\ -17 \\ \hline \end{array}$	$\begin{array}{r} 64 \\ -37 \\ \hline \end{array}$	$\begin{array}{r} 70 \\ -26 \\ \hline \end{array}$
3.	$\begin{array}{r} 64 \\ -37 \\ \hline \end{array}$	$\begin{array}{r} 84 \\ -25 \\ \hline \end{array}$	$\begin{array}{r} 42 \\ -24 \\ \hline \end{array}$	$\begin{array}{r} 78 \\ -39 \\ \hline \end{array}$	$\begin{array}{r} 47 \\ -38 \\ \hline \end{array}$	$\begin{array}{r} 63 \\ -47 \\ \hline \end{array}$
4.	$\begin{array}{r} 153 \\ -27 \\ \hline \end{array}$	$\begin{array}{r} 625 \\ -16 \\ \hline \end{array}$	$\begin{array}{r} 171 \\ -25 \\ \hline \end{array}$	$\begin{array}{r} 835 \\ -19 \\ \hline \end{array}$	$\begin{array}{r} 135 \\ -16 \\ \hline \end{array}$	$\begin{array}{r} 635 \\ -26 \\ \hline \end{array}$
5.	$\begin{array}{r} 543 \\ -26 \\ \hline \end{array}$	$\begin{array}{r} 175 \\ -28 \\ \hline \end{array}$	$\begin{array}{r} 483 \\ -75 \\ \hline \end{array}$	$\begin{array}{r} 125 \\ -19 \\ \hline \end{array}$	$\begin{array}{r} 785 \\ -49 \\ \hline \end{array}$	$\begin{array}{r} 135 \\ -27 \\ \hline \end{array}$
6.	$\begin{array}{r} 524 \\ -15 \\ \hline \end{array}$	$\begin{array}{r} 684 \\ -25 \\ \hline \end{array}$	$\begin{array}{r} 895 \\ -26 \\ \hline \end{array}$	$\begin{array}{r} 768 \\ -29 \\ \hline \end{array}$	$\begin{array}{r} 726 \\ -17 \\ \hline \end{array}$	$\begin{array}{r} 325 \\ -16 \\ \hline \end{array}$
7.	$\begin{array}{r} 385 \\ -76 \\ \hline \end{array}$	$\begin{array}{r} 735 \\ -17 \\ \hline \end{array}$	$\begin{array}{r} 633 \\ -15 \\ \hline \end{array}$	$\begin{array}{r} 354 \\ -45 \\ \hline \end{array}$	$\begin{array}{r} 576 \\ -47 \\ \hline \end{array}$	$\begin{array}{r} 880 \\ -35 \\ \hline \end{array}$

Lesson 4   Problem Solving

Solve each problem.

1. A candle was 32 centimeters long when it was lit. When it was blown out, it was 19 centimeters long. How long was the part that burned away?

   The candle was _____ centimeters long when it was lit.

   After burning, it was _____ centimeters long.

   _____ centimeters of the candle burned away.

2. A gasoline tank can hold 92 liters. It took 68 liters to fill the tank. How many liters were in the tank before filling it?

   The tank can hold _____ liters.

   It took _____ liters to fill the tank.

   _____ liters were in the tank before filling.

3. Joy exercised 42 minutes this morning. She exercised 27 minutes this afternoon. How many more minutes did she exercise this morning?

   Joy exercised _____ minutes this morning.

   She exercised _____ minutes this afternoon.

   Joy exercised _____ minutes more this morning.

4. There are 283 students at Adams School. Sixty-seven of these students were in the park program last year. How many were not in the program?

   _____ students were not in the program.

5. Pablo's father weighs 173 pounds. Pablo weighs 65 pounds. How much more does Pablo's father weigh?

   Pablo's father weighs _____ pounds more.

1.

2.

3.

4.   **5.**

Lesson 5 Subtraction (2- and 3-digit)

	Rename 306 as "2 hundreds, 10 tens, and 6 ones."	Then rename as "2 hundreds, 9 tens, and 16 ones." Subtract the ones.	Subtract the tens.	Subtract the hundreds.

```
  306        2 10         9             9             9
             3 0 6       2 10 16       2 10 16       2 10 16
 -89         3 0 6       3 0 6         3 0 6         3 0 6
            -8 9        -8 9          -8 9          -8 9
                              7          1 7         2 1 7
```

Subtract.

	a	b	c	d	e	f
1.	154 −61	125 −74	107 −56	137 −50	124 −53	126 −71
2.	411 −80	429 −39	862 −91	428 −37	784 −93	605 −71
3.	984 −97	352 −65	604 −75	671 −93	325 −46	864 −79
4.	168 −69	906 −97	384 −96	763 −94	286 −98	857 −69
5.	666 −88	127 −49	700 −99	318 −79	722 −59	821 −48
6.	454 −67	357 −89	183 −94	306 −58	854 −95	347 −79
7.	346 −88	464 −87	207 −49	123 −34	253 −78	241 −59

Lesson 5   Problem Solving

Solve each problem.

1. A school building is 64 feet high. A flagpole is 125 feet high. How much higher is the flagpole?

    The flagpole is _____ feet high.

    The school building is _____ feet high.

    The flagpole is _____ feet higher.

2. Steve threw a ball 117 feet. His cousin threw it 86 feet. How much farther did Steve throw it?

    Steve threw the ball _____ feet.

    His cousin threw the ball _____ feet.

    Steve threw the ball _____ feet farther.

3. Eight hundred sixty-five tickets went on sale for a concert. So far 95 tickets have been sold. How many tickets are left?

    _____ tickets went on sale.

    _____ tickets have been sold.

    _____ tickets are left.

4. Last year Mrs. Moore's bowling average was 91. This year her average is 123. How much has her bowling average improved over last year?

    Her average has improved _____ points.

5. Brittany is reading a 234-page book. She has read 57 pages. How many more pages does she still have to read?

    She has _____ pages yet to read.

1.

2.

3.

4.   **5.**

Lesson 6 Checking Addition and Subtraction

To check 62 + 57 = 119, subtract 57 from 119.

$$\begin{array}{r} 62 \\ +57 \\ \hline 119 \\ -57 \\ \hline 62 \end{array}$$

These should be the same.

To check 125 − 67 = 58, add 67 to 58.

$$\begin{array}{r} 125 \\ -67 \\ \hline 58 \\ +67 \\ \hline 125 \end{array}$$

These should be the same.

Add. Check each answer.

	a	b	c	d	e	f
1.	42 +25	63 +35	24 +64	26 +47	38 +42	59 +35
2.	63 +45	73 +56	83 +35	78 +25	62 +89	98 +95

Subtract. Check each answer.

3.	85 −24	79 −45	68 −39	95 −28	68 −29	73 −48
4.	125 −63	146 −83	164 −73	104 −86	152 −64	186 −97

Lesson 7   Addition and Subtraction Review

Add.

	a	*b*	*c*	*d*	*e*	*f*
1.	5 7 +9	3 5 +7	9 8 +6	5 3 +8	6 7 +5	4 3 +7
2.	38 +6	47 +9	63 +8	7 +49	8 +67	9 +47
3.	47 +24	43 +48	91 +26	46 +74	28 +356	435 +93
4.	15 22 +45	27 33 +18	31 12 +85	31 87 +41	17 156 +42	785 31 +56

Subtract.

	a	*b*	*c*	*d*	*e*	*f*
5.	36 −7	48 −9	63 −8	45 −7	54 −8	76 −9
6.	43 −25	75 −18	93 −44	50 −17	71 −28	67 −49
7.	147 −29	264 −25	784 −39	673 −54	270 −53	687 −59
8.	245 −93	705 −73	248 −75	638 −85	459 −87	317 −45
9.	345 −76	508 −59	867 −89	316 −39	707 −48	465 −68

CHAPTER 2 PRACTICE TEST
Addition and Subtraction (2- and 3-digit; with renaming)

Add. Check each answer.

	a	b	c	d	e	f
1.	37 +6	8 +47	59 +8	5 +48	63 +9	9 +81

2.	65 +27	34 +92	88 +37	159 +82	267 +76	347 +96

Subtract. Check each answer.

3.	84 −27	74 −45	68 −49	93 −38	61 −47	78 −59

4.	135 −64	126 −84	463 −72	153 −96	384 −96	302 −85

Solve.

5. Emily weighs 48 kilograms, Joshua weighs 33 kilograms, and Andrew weighs 37 kilograms. What is their combined weight?

5.

Their combined weight is _____ kilograms.

CHAPTER 3 PRETEST
Addition and Subtraction (3-digit through 5-digit)

Add.

	a	*b*	*c*	*d*	*e*	*f*
1.	621 +214	345 +332	426 +153	425 +316	245 +127	458 +329
2.	425 +193	373 +282	625 +193	624 +732	506 +792	591 +805
3.	397 +113	287 +125	926 +7287	3452 +1139	4646 +1283	5252 +3934
4.	31 10 +24	25 16 +23	232 151 +474	531 612 +743	2137 3272 +1324	4573 2281 +1654
5.	12 21 31 +12	200 413 134 +131	3171 1540 2134 +1023	3421 1313 1510 +2643	3117 2375 1132 +1214	3774 1571 3232 +1351

Subtract.

	a	*b*	*c*	*d*	*e*	*f*
6.	657 −234	745 −416	967 −173	406 −257	5627 −512	4848 −425
7.	4357 −138	5725 −273	6753 −902	7425 −286	8652 −937	6053 −782
8.	4357 −1132	5678 −1429	3675 −1294	5678 −2923	7802 −3254	9797 −1898
9.	67524 −1321	34723 −2308	78243 −4152	80145 −1913	76762 −9341	78545 −2837

Lesson 1 Addition and Subtraction (3- and 4-digit)

Add the ones.	Add the tens.	Add the hundreds.	Subtract the ones.	Rename and subtract the tens.	Rename and subtract the hundreds.
783 +562 ⎯⎯ 5	¹ 783 +562 ⎯⎯ 45	¹ 783 +562 ⎯⎯ 1345	1253 −582 ⎯⎯ 1	1 15 1 2̸ 5̸ 3 −5 8 2 ⎯⎯ 7 1	11 1̸ 15 1̸ 2̸ 5̸ 3 −5 8 2 ⎯⎯ 6 7 1

CHAPTER 3

Add.

	a	b	c	d	e	f
1.	4 3 2 +3 2 5	3 2 5 +5 3 6	2 3 2 +5 7 3	3 2 5 +8 1 4	3 2 8 +1 9 3	5 2 9 +2 8 7
2.	6 7 5 +9 0 7	8 6 7 +3 2 5	6 2 5 +5 9 4	8 9 1 +5 3 6	6 7 5 +7 3 8	4 7 5 +9 6 9
3.	3 5 7 +5 2 8	1 4 6 +4 9 4	7 3 4 +8 5 9	5 3 6 +6 7 3	8 6 7 +7 9 5	8 9 3 +7 5 7

Subtract.

	a	b	c	d	e
4.	8 5 7 −1 4 3	7 5 7 −1 2 9	4 6 7 −1 8 2	9 5 2 −2 7 8	8 6 3 −3 8 9
5.	1 4 8 9 −5 2 7	1 2 4 6 −8 1 3	1 5 7 8 −9 2 7	1 7 2 8 −9 1 9	1 3 7 3 −5 4 8
6.	1 4 5 8 −7 7 3	1 7 3 2 −9 6 1	1 4 2 0 −7 3 5	1 2 5 4 −9 9 5	1 6 5 2 −7 9 7
7.	1 0 2 0 −5 5 2	1 2 3 4 −4 3 5	1 6 0 0 −9 0 0	1 3 5 7 −9 6 8	1 0 0 0 −8 7 9

Lesson 1 Problem Solving

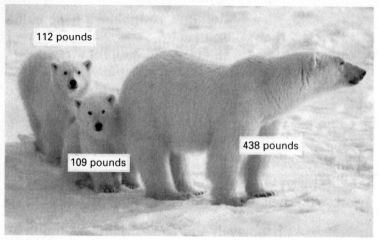

112 pounds

109 pounds

438 pounds

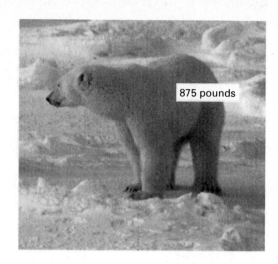

875 pounds

Solve each problem.

1. Find the combined weight of the two smallest polar bears.

   Are you to add or subtract? _____

   The combined weight is _____ pounds.

   1.

2. How much more does the largest bear weigh than the smallest bear?

   Are you to add or subtract? _____

   The largest bear weighs _____ pounds more than the smallest bear.

   2.

3. The adult male bear is 118 inches long. The adult female bear is 69 inches long. How much longer is the male than the female?

   Are you to add or subtract? _____

   The male bear is _____ inches longer than the female bear.

   3.

4. Find the combined weight of the two largest polar bears.

   Are you to add or subtract? _____

   The combined weight is _____ pounds.

   4.

Lesson 2 Addition and Subtraction (4- and 5-digit)

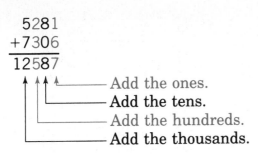

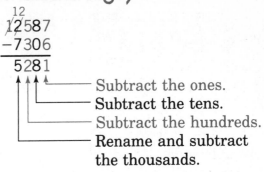

```
  5281            ₁₂
 +7306          1̶2̶587
 ─────          −7306
 12587          ─────
                 5281
```

—— Add the ones.
—— Add the tens.
—— Add the hundreds.
—— Add the thousands.

—— Subtract the ones.
—— Subtract the tens.
—— Subtract the hundreds.
—— Rename and subtract
  the thousands.

CHAPTER 3

Add.

	a	*b*	*c*	*d*	*e*
1.	3 2 5 3 +1 4 2 4	6 7 2 5 +1 1 3 8	2 4 5 6 +3 2 9 3	5 2 6 3 +1 8 2 4	6 2 4 5 +7 4 0 3
2.	5 4 7 8 +2 3 2 5	5 7 6 9 +3 9 1 4	6 5 1 8 +9 2 4 6	5 3 8 2 +1 6 6 4	7 6 8 3 +8 1 8 5
3.	8 1 2 1 +7 9 3 4	2 5 2 3 +4 6 9 4	6 3 7 8 +7 1 6 4	3 2 3 5 +4 9 1 7	8 4 8 3 +9 6 5 8

Subtract.

	a	*b*	*c*	*d*	*e*
4.	2 6 7 5 −4 3 6	3 6 7 5 −5 9 4	6 7 2 5 −9 0 4	4 3 0 7 −2 8 9	6 3 5 2 −7 6 4
5.	4 3 5 7 −2 3 4 9	6 7 3 5 −1 2 6 4	9 5 2 5 −4 6 0 3	5 6 7 5 −2 3 8 9	6 0 5 2 −2 9 6 8
6.	1 6 7 8 4 −7 3 2 5	1 2 5 4 3 −3 3 6 2	1 5 7 4 7 −6 9 3 6	1 0 1 3 7 −9 6 5 2	1 7 6 7 5 −8 8 9 6
7.	1 4 3 2 1 −3 4 2 1	1 0 2 3 6 −1 5 2 7	1 3 5 0 0 −6 8 7 0	1 1 2 3 8 −4 7 3 9	1 0 0 0 0 −6 7 8 2

Lesson 2 Problem Solving

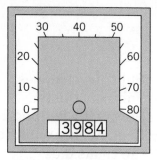

Mrs. Crannell's car

Mr. Comm's car

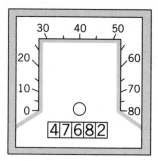

Mr. Hsu's car

Solve each problem.

1. Mr. Hsu's car has been driven how many more miles than Mr. Comm's car?

   Mr. Hsu's car has been driven _____ miles.

   Mr. Comm's car has been driven _____ miles.

   Mr. Hsu's car has been driven _____ miles more.

2. Mr. Hsu plans to drive his car 870 miles on a trip. How many miles will be on his car after the trip?

   _____ miles are on his car now.

   The trip will be _____ miles long.

   _____ miles will be on his car after the trip.

3. Mrs. Crannell's car needs an oil change at 6,000 miles. How many more miles can she go before changing oil?

   She can go _____ more miles.

4. Mr. Mitchel wants to buy a used van for $8,365 or a car for $6,978. How much more does the van cost?

   The van costs $ _____ more.

5. St. Louis is 1,933 kilometers from Boston. San Francisco is 3,414 kilometers from St. Louis. How far is it from Boston to San Francisco by way of St. Louis?

   The distance is _____ kilometers.

1.

2.

3.

4.

5.

Lesson 3   Addition (three or more numbers)

Add in each place-value position from right to left.

```
                                      122
                        212          4325
          1 1          1632          6078
         6374          3779          5298
        +2809         +6809         +5764
         9183         12220         21465
```

Add.

	a	b	c	d	e

1.
```
    2 4      4 6      4 5      4 5      5 2
    3 1      2 3      6 2      1 3      2 3
   +4 0     +1 5     +7 1     +2 1     +7 1
```

2.
```
    3 4      1 2 6      3 4 5      5 2 4      3 0 5
    2 1      1 2        1 6 2      6 3 0      1 3 1
  +1 1 2    +6 2 4      +7 1      +7 2 1     +4 2 2
```

3.
```
  3 2 3 5    2 1 4 5    8 2 1 8    1 3 5 3    4 4 3 5
  3 1 1 2    3 4 1 8    3 2 4 5    2 3 3 1    8 2 7 1
 +1 4 8 6   +1 9 3 2   +4 1 2 3   +3 6 4 2   +4 1 6 0
```

4.
```
  5 6 4 1    1 8 2 6    7 1 3 7    2 4 5 3    3 4 1 7
  2 7 2 2    2 5 7 4    8 0 2 8    8 7 4 2    8 7 0 3
 +4 8 3 3   +4 4 9 3   +7 6 5 6   +2 5 6 1   +2 8 5 4
```

5.
```
  5 2 4 7    3 2 5 3    1 6 0 1    5 1 4 5    1 0 1 1
  2 4 0 3    1 1 6 1    2 7 2 2    6 2 0 1    2 4 6 2
  1 1 2 5    1 1 7 2    3 8 1 3    2 3 1 2    3 5 7 1
 +1 0 1 7   +4 0 8 0   +1 2 4 1   +4 0 2 1   +1 2 5 4
```

6.
```
  1 0 2 5    1 5 4 6    4 1 2 4    3 6 5 2    6 3 1 7
  3 1 1 3    2 3 3 5    1 2 3 1    6 2 7 4    2 1 6 4
  1 2 5 8    3 8 2 2    5 3 5 2    3 1 7 5    5 5 7 3
 +2 4 6 4   +1 9 4 1   +6 0 7 5   +5 1 1 2   +4 2 5 8
```

Lesson 3   Problem Solving

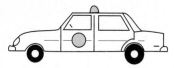   Dot Cab Company
September Report

Cab number	Liters of gasoline used	Number of hours driven	Number of kilometers traveled	Number of passengers	Amount of fares collected
100	707	566	6,307	2,205	$8,221
101	791	573	6,962	2,542	$8,825
102	729	548	6,566	1,821	$8,533

Solve each problem.

1. How many liters of gasoline were used by the three cabs during the month of September?

_____ liters were used by cab 100.

_____ liters were used by cab 101.

_____ liters were used by cab 102.

_____ liters were used in all.

1.

2. How many hours were the three cabs driven?

The cabs were driven _____ hours.

3. How many passengers rode in the three cabs?

_____ passengers rode in the cabs.

2.

3.

4. How many kilometers did the three cabs travel?

The cabs traveled _____ kilometers.

5. How much was collected in fares for the three cabs?

$ _____ was collected in fares.

4.

5.

Lesson 4 Addition and Subtraction Review

Add.

	a	*b*	*c*	*d*	*e*
1.	312 +541	135 +427	231 +384	532 +614	537 +148
2.	1456 +218	295 +2461	723 +5413	4561 +349	4624 +597
3.	5247 +2216	4673 +4285	4672 +1401	5314 +7053	4314 +3197
4.	42 13 +25	131 12 +449	213 227 +384	4163 5274 +6521	4712 1824 +5531
5.	67 78 55 +27	731 142 253 +461	7325 2106 7347 +2511	5314 6024 7151 +2235	5678 2345 6789 +4257

Subtract.

	a	*b*	*c*	*d*	*e*
6.	687 −434	754 −236	576 −393	605 −388	795 −498
7.	1234 −125	3857 −665	4257 −843	4657 −839	5014 −968
8.	7354 −4038	5619 −2348	4187 −2574	6753 −1942	7815 −4176
9.	42573 −1846	36154 −9038	46124 −9762	54751 −2896	70534 −7689

CHAPTER 3

Lesson 4 Problem Solving

Solve each problem.

1. An empty truck weighs 1,750 kilograms. It is to be loaded with 1,402 kilograms of cargo. What will be the weight of the truck and its cargo?

   The truck weighs _____ kilograms.

   The cargo weighs _____ kilograms.

   The combined weight is _____ kilograms.

2. On Monday, 9,852 fans attended the game. There were 7,569 fans at the game on Tuesday. How many fans attended the two games?

   _____ fans attended the games.

3. Ms. Krom has saved $10,320 for a new car. Her goal is to save $12,500. How much more must she save?

   She must save $_____ more.

4. Complete the table.

5. How many more women are there than men?

   There are

   _____ more women.

Benny School	
Number of women	1,029
Number of men	983
Total	

6. Complete the table.

7. How many more votes did Witt get than Lewis?

   Witt got

   _____ more votes.

Election results	*Number of votes*
Witt	4,327
Lewis	2,539
Total	

1.

2.

3.

5.

7.

Lesson 5 Rounding

Round 2,357 to the nearest hundred. Look at the tens digit.	Round 12,345 to the nearest thousand. Look at the hundreds digit.
2,357	12,345
5 is greater than or equal to 5, so round 3 to 4 in the hundreds place. Follow with zeros.	3 is less than 5, so keep 2 in the thousands place and follow with zeros.
2,400	12,000

Round to the nearest ten.

	a	b	c	d	e
1.	123	21,137	149	5,115	81,172
	_____	_____	_____	_____	_____
2.	103	7,196	10,154	125	88
	_____	_____	_____	_____	_____

Round to the nearest hundred.

	a	b	c	d	e
3.	36,592	21,504	238	5,679	98
	_____	_____	_____	_____	_____
4.	4,852	22,359	563	12,053	38,245
	_____	_____	_____	_____	_____

Round to the nearest thousand.

	a	b	c	d	e
5.	5,729	12,356	34,722	53,098	585
	_____	_____	_____	_____	_____
6.	17,526	4,892	23,574	9,256	85,236
	_____	_____	_____	_____	_____

Round to the nearest ten-thousand.

	a	b	c	d	e
7.	36,579	9,800	52,693	67,532	25,374
	_____	_____	_____	_____	_____
8.	85,320	22,568	94,268	55,555	12,105
	_____	_____	_____	_____	_____

Lesson 5 Problem Solving

Solve each problem.

1. The distance from Seattle, Washington, to Albany, New York, is 2,209 miles.

How far apart to the nearest hundred miles are these two cities?

_____ miles

How far to the nearest thousand miles are they apart?

_____ miles

2. Yolanda bought her first home for $87,546.

How much to the nearest hundred dollars did she pay?

_____

How much to the nearest thousand dollars did her home cost?

_____

What to the nearest ten-thousand dollars did she pay?

_____

3. Jacob had a collection of 562 trading cards.

To the nearest ten cards, how many cards had he collected?

_____ cards

How many cards to the nearest hundred cards did he have?

_____ cards

4. Mt. Everest is 29,028 feet tall. It is the highest mountain in the world.

How tall to the nearest thousand feet is Mt. Everest?

_____ feet

How tall to the nearest ten-thousand feet is Mt. Everest?

_____ feet

1.

2.

3.

4.

Lesson 6 Estimation (sums and differences)

Round each number to the highest place value the numbers have in common. Then add from right to left.

$$1276 \longrightarrow 1300$$
$$+ 354 \longrightarrow +400$$
$$1,700$$

Round each number to the highest place value the numbers have in common. Then subtract from right to left.

$$8486 \longrightarrow 8000$$
$$+4729 \longrightarrow +5000$$
$$3,000$$

Estimate each sum or difference.

	a	b	c	d	e
1.	594 +236	259 +767	636 +158	236 +269	2776 +9278
2.	679 −256	868 −289	894 −643	716 −359	2974 −1629
3.	2774 +364	7671 +795	8235 +4691	6752 +462	46713 +1125
4.	886 −567	1574 −679	6152 −364	5695 −262	3115 −645
5.	5176 +859	33495 +2850	62954 +12795	34592 +866	21811 +6495
6.	65902 −45792	98410 +18642	63379 −4451	33285 −123	21362 −556

Lesson 6 Problem Solving

Answer each question. Use estimation.

1. A mother elephant weighed 13,560 pounds. Her newborn baby weighed 220 pounds. About how much more did the mother weigh than the baby?

   Do you add or subtract? _____

   About how much more did the mother weigh than the baby?

   about _____ pounds

   1.

2. On Saturday, 10,230 people came to the city festival. On Sunday, it was raining, so only 880 people attended. About how many people came to the weekend festival?

   Do you add or subtract? _____

   About how many people came to the weekend festival?

   about _____ people

   2.

3. Thomas has $356 in his checking account and $4,225 in his savings account. About how much money does Thomas have in all?

   Do you add or subtract? _____

   About how much money does Thomas have in all?

   about _____

   3.

4. Maria's class sold $1,220 worth of magazines for their school, and Jose's class sold $985 for their school. How much more was sold in Maria's class than in Jose's class?

   Do you add or subtract? _____

   How much more was sold in Maria's class than in Jose's class?

   about _____

   4.

CHAPTER 3 PRACTICE TEST
Addition and Subtraction (3-digit through 5-digit)

Add or subtract.

	a	b	c	d
1.	2 1 2 +1 6 5	5 2 5 +1 3 6	4 5 2 1 1 4 +1 2 0	1 2 1 1 1 6 5 3 2 +4 9 5 1
2.	6 2 5 − 4 0 3	8 2 5 − 1 3 2	1 6 0 8 5 − 1 8 0 4	5 4 2 1 3 − 9 9 8 8

Round each number to the given place.

	a	b	c
3.	157; tens _____	51,238; thousands _____	5,121; hundreds _____
4.	1,679; hundreds _____	93,220; ten-thousands _____	3,198; thousands _____

Estimate each sum or difference.

	a	b	c	d
5.	6 9 4 +1 3 5	2 3 6 +3 5 9	1 6 7 1 3 +3 1 1 2 0	6 8 8 5 + 2 3 6
6.	8 9 8 − 2 5 6	9 5 4 − 6 7 9	3 6 2 5 − 7 5 4	3 2 9 7 4 − 1 1 6 2 9

7. The Smith family drove 1,050 miles from Dallas, Texas, to Columbus, Ohio. Then they drove 619 miles from Columbus to Albany, New York. About how many total miles did the Smith family drive?

7.

The Smith family drove about _____ miles.

CHAPTER 3

CHAPTER 4 PRETEST
Multiplication (2- and 3-digit by 1-digit)

	a	*b*	*c*	*d*	*e*	*f*
1.	7 ×4	6 ×9	8 ×5	40 ×2	20 ×3	10 ×8
2.	23 ×2	21 ×4	32 ×3	11 ×5	21 ×3	22 ×2
3.	22 ×3	21 ×2	43 ×2	23 ×3	22 ×4	34 ×2
4.	17 ×5	27 ×3	15 ×6	23 ×4	12 ×7	12 ×8
5.	51 ×9	62 ×4	71 ×8	67 ×7	96 ×3	83 ×5
6.	200 ×4	300 ×2	400 ×2	121 ×4	124 ×2	312 ×3
7.	105 ×9	124 ×4	325 ×3	121 ×8	172 ×4	283 ×3
8.	400 ×7	412 ×3	924 ×4	513 ×7	618 ×4	176 ×5
9.	830 ×7	731 ×6	138 ×7	673 ×8	469 ×5	869 ×4

Lesson 1 Multiplication Facts

5 → Find the **5** -row.

× 6 → Find the **6** -column.

30 ← The product is named where the 5-row and 6-column meet.

Use the table to multiply.

7
× 9
—

6-column

×	0	1	2	3	4	5	6	7	8	9
0	0	0	0	0	0	0	0	0	0	0
1	0	1	2	3	4	5	6	7	8	9
2	0	2	4	6	8	10	12	14	16	18
3	0	3	6	9	12	15	18	21	24	27
4	0	4	8	12	16	20	24	28	32	36
5	0	5	10	15	20	25	30	35	40	45
6	0	6	12	18	24	30	36	42	48	54
7	0	7	14	21	28	35	42	49	56	63
8	0	8	16	24	32	40	48	56	64	72
9	0	9	18	27	36	45	54	63	72	81

5-row

CHAPTER 4

Multiply.

	a	b	c	d	e	f	g	h
1.	6 ×1	7 ×8	8 ×9	9 ×0	7 ×7	6 ×7	7 ×1	4 ×6
2.	9 ×9	6 ×2	7 ×6	9 ×1	8 ×1	6 ×8	8 ×8	3 ×7
3.	8 ×0	7 ×5	6 ×0	9 ×2	8 ×7	6 ×9	7 ×0	4 ×7
4.	5 ×6	9 ×3	7 ×4	6 ×3	9 ×4	4 ×8	8 ×6	3 ×6
5.	7 ×9	5 ×7	5 ×8	8 ×5	6 ×4	7 ×3	9 ×5	2 ×6
6.	7 ×2	8 ×4	9 ×6	9 ×7	4 ×9	6 ×5	9 ×8	2 ×8
7.	8 ×3	5 ×9	3 ×8	8 ×2	3 ×9	2 ×7	6 ×6	2 ×9

Lesson 1   Problem Solving

Solve each problem.

1. Last week Sean's father worked five 8-hour shifts. How many hours did he work last week?

   He worked _____ shifts.

   There were _____ hours in each shift.

   He worked _____ hours last week.

2. A certain factory operates two 8-hour shifts each day. How many hours does the factory operate each day?

   There are _____ shifts.

   There are _____ hours in each shift.

   The factory operates _____ hours each day.

3. It takes the cleanup crew four hours to clean the factory after each day's work. How many hours will the cleanup crew work during a five-day week?

   The cleanup crew works _____ hours a day.

   They work _____ days a week.

   The cleanup crew works _____ hours a week.

4. Lauren's mother works five hours each day. She works five days each week. How many hours does she work each week?

   She works _____ hours each week.

5. It costs a company $8 an hour to operate a certain machine. How much will it cost to operate the machine for six hours?

   It will cost $_____.

1.		
2.		
3.		
4.		5.

Lesson 2 Multiplication (2-digit by 1-digit)

```
   4        40              43      Multiply      Multiply
  ×2        ×2              ×2      3 ones by 2.  4 tens by 2.
  ──        ──                        43            43
   8        80                       ×2            ×2
                                     ──            ──
If 2 × 4 = 8,                         6            86
then 2 × 40 = 80.
```

Multiply.

	a	b	c	d	e	f
1.	3 ×3	30 ×3	2 ×4	20 ×4	3 ×2	30 ×2
2.	2 ×3	30 ×3	32 ×3	1 ×2	40 ×2	41 ×2
3.	11 ×9	33 ×3	12 ×3	14 ×2	31 ×3	13 ×3
4.	32 ×2	23 ×3	42 ×2	21 ×4	13 ×2	11 ×6
5.	12 ×2	11 ×5	33 ×2	11 ×3	21 ×2	22 ×3
6.	11 ×4	44 ×2	22 ×2	11 ×8	11 ×2	13 ×2
7.	23 ×2	22 ×4	24 ×2	21 ×3	31 ×2	11 ×7

Lesson 2   Problem Solving

Solve each problem.

1. There are 1 dozen cans of peaches in each carton. How many cans are in two cartons? Remember, there are 12 items in 1 dozen.

   _____ cans are in 1 dozen.

   There are _____ cartons.

   There are _____ cans of peaches in two cartons.

2. Twelve cans of pineapple are in each carton. How many cans are in three cartons?

   There are _____ cans of pineapple in three cartons.

3. There are 1 dozen cans of pears in each carton. How many cans are in four cartons?

   There are _____ cans of pears in four cartons.

4. Ten cans of orange sections come in each carton. How many cans are in four cartons?

   There are _____ cans of orange sections in four cartons.

1.	2.
3.	4.

Lesson 3 Multiplication (2-digit by 1-digit; renaming)

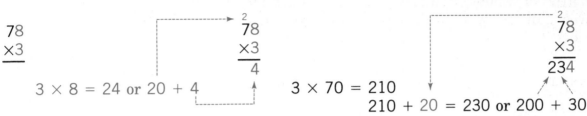

Multiply
8 ones by 3.

Multiply 7 tens by 3.
Add the 2 tens.

$$
\begin{array}{r} 78 \\ \times 3 \\ \hline \end{array}
\qquad
\begin{array}{r} {}^{2}\\ 78 \\ \times 3 \\ \hline 4 \end{array}
\qquad
\begin{array}{r} {}^{2}\\ 78 \\ \times 3 \\ \hline 234 \end{array}
$$

$3 \times 8 = 24$ or $20 + 4$ $3 \times 70 = 210$

$210 + 20 = 230$ or $200 + 30$

Multiply.

	a	*b*	*c*	*d*	*e*	*f*
1.	$\begin{array}{r}28\\ \times2\\ \hline\end{array}$	$\begin{array}{r}23\\ \times4\\ \hline\end{array}$	$\begin{array}{r}25\\ \times3\\ \hline\end{array}$	$\begin{array}{r}16\\ \times6\\ \hline\end{array}$	$\begin{array}{r}19\\ \times5\\ \hline\end{array}$	$\begin{array}{r}37\\ \times2\\ \hline\end{array}$
2.	$\begin{array}{r}14\\ \times6\\ \hline\end{array}$	$\begin{array}{r}13\\ \times7\\ \hline\end{array}$	$\begin{array}{r}29\\ \times3\\ \hline\end{array}$	$\begin{array}{r}12\\ \times8\\ \hline\end{array}$	$\begin{array}{r}46\\ \times2\\ \hline\end{array}$	$\begin{array}{r}12\\ \times7\\ \hline\end{array}$
3.	$\begin{array}{r}53\\ \times3\\ \hline\end{array}$	$\begin{array}{r}61\\ \times5\\ \hline\end{array}$	$\begin{array}{r}74\\ \times2\\ \hline\end{array}$	$\begin{array}{r}81\\ \times6\\ \hline\end{array}$	$\begin{array}{r}71\\ \times7\\ \hline\end{array}$	$\begin{array}{r}62\\ \times4\\ \hline\end{array}$
4.	$\begin{array}{r}92\\ \times4\\ \hline\end{array}$	$\begin{array}{r}73\\ \times2\\ \hline\end{array}$	$\begin{array}{r}91\\ \times5\\ \hline\end{array}$	$\begin{array}{r}61\\ \times7\\ \hline\end{array}$	$\begin{array}{r}72\\ \times3\\ \hline\end{array}$	$\begin{array}{r}61\\ \times8\\ \hline\end{array}$
5.	$\begin{array}{r}73\\ \times9\\ \hline\end{array}$	$\begin{array}{r}85\\ \times2\\ \hline\end{array}$	$\begin{array}{r}59\\ \times6\\ \hline\end{array}$	$\begin{array}{r}48\\ \times7\\ \hline\end{array}$	$\begin{array}{r}67\\ \times3\\ \hline\end{array}$	$\begin{array}{r}57\\ \times8\\ \hline\end{array}$
6.	$\begin{array}{r}72\\ \times6\\ \hline\end{array}$	$\begin{array}{r}76\\ \times4\\ \hline\end{array}$	$\begin{array}{r}83\\ \times9\\ \hline\end{array}$	$\begin{array}{r}98\\ \times5\\ \hline\end{array}$	$\begin{array}{r}54\\ \times7\\ \hline\end{array}$	$\begin{array}{r}42\\ \times9\\ \hline\end{array}$
7.	$\begin{array}{r}83\\ \times5\\ \hline\end{array}$	$\begin{array}{r}74\\ \times6\\ \hline\end{array}$	$\begin{array}{r}97\\ \times9\\ \hline\end{array}$	$\begin{array}{r}58\\ \times8\\ \hline\end{array}$	$\begin{array}{r}74\\ \times7\\ \hline\end{array}$	$\begin{array}{r}49\\ \times4\\ \hline\end{array}$

Lesson 3 Problem Solving

Solve each problem.

1. A bottle of Sudsy Shampoo contains 12 fluid ounces. How many fluid ounces are in three such bottles?

There are _____ fluid ounces in each bottle.

There are _____ bottles.

There are _____ fluid ounces in three bottles.

2. Mr. Long drives 14 miles to work each day. He works six days a week. How far does he drive to work each week?

He drives _____ miles each day.

He works _____ days each week.

He drives _____ miles each week.

3. A case contains 24 cans. How many cans will be in nine such cases?

Each case contains _____ cans.

There are _____ cases.

There are _____ cans in nine cases.

4. The Acme Salt Company shipped eight sacks of salt to the Sour Pickle Company. Each sack of salt weighed 72 pounds. What was the weight of the shipment?

The total weight of the shipment was

_____ pounds.

5. A train can travel 69 kilometers in one hour. How far can it travel in four hours?

It can travel _____ kilometers.

1.

2.

3.

4.

5.

Lesson 4   Multiplication (3-digit by 1-digit; renaming)

| Multiply 2 ones by 4. | Multiply 8 tens by 4. | Multiply 6 hundreds by 4. Add the 3 hundreds. |

$$\begin{array}{r} 4 \\ \times 6 \\ \hline 24 \end{array} \qquad \begin{array}{r} 400 \\ \times 6 \\ \hline 2400 \end{array}$$

$$\begin{array}{r} 682 \\ \times 4 \\ \hline 8 \end{array}$$

$$4 \times 80 = 320$$

$$320 = 300 + 20$$

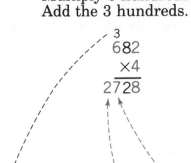

$$4 \times 600 = 2400$$
$$2400 + 300 = 2700 = 2000 + 700$$

Multiply.

	a	*b*	*c*	*d*	*e*	*f*
1.	3 ×2	3 0 0 ×2	5 ×7	5 0 0 ×7	6 ×8	6 0 0 ×8
2.	1 8 × 4	2 0 0 ×4	2 1 8 ×4	2 1 ×4	7 0 0 ×4	7 2 1 ×4

Multiply.

	a	*b*	*c*	*d*	*e*
3.	3 2 1 ×3	4 2 3 ×2	2 1 2 ×4	3 4 9 ×2	3 2 7 ×3
4.	1 5 6 ×6	2 3 8 ×4	8 0 7 ×6	4 1 3 ×7	4 2 1 ×8
5.	9 8 7 ×6	4 7 8 ×9	6 7 8 ×7	7 2 7 ×8	5 9 4 ×5

Lesson 4 Problem Solving

Solve each problem.

1. An airplane can carry 183 passengers. How many passengers could five such airplanes carry?

   They could carry _____ passengers.

   1.

2. The school ordered seven sets of books. There are 125 books in each set. How many books were ordered?

   _____ books were ordered.

   2.

3. Each family in a building was given four keys. There were 263 families in the building. How many keys were given out?

   _____ keys were given out.

   3.

4. Each student receives two cartons of milk a day. There are 912 students in the school. How many cartons of milk will be needed?

   _____ cartons of milk will be needed.

   4.

5. There are 217 apartments in Lindsay's building. Each apartment has six windows. How many windows are there in all?

   There are _____ windows in all.

   5.

6. Tusk the elephant eats 145 pounds of food a day. How many pounds of food will he eat in seven days?

   Tusk will eat _____ pounds of food in seven days.

   6.

7. Carlos delivers 128 papers each day. How many papers will he deliver in six days?

   He will deliver _____ papers in six days.

   7.

Lesson 5 Multiplication Review

Multiply.

	a	b	c	d	e	f
1.	11 ×7	32 ×2	21 ×4	32 ×3	24 ×2	23 ×3
2.	13 ×7	15 ×5	47 ×2	13 ×6	24 ×4	25 ×3
3.	72 ×3	92 ×4	71 ×6	62 ×4	81 ×5	93 ×3
4.	66 ×4	48 ×7	67 ×6	72 ×9	57 ×7	49 ×5

Multiply.

	a	b	c	d	e
5.	214 ×2	231 ×3	210 ×4	224 ×4	115 ×6
6.	117 ×5	131 ×7	394 ×2	121 ×8	732 ×3
7.	912 ×4	610 ×5	157 ×6	235 ×4	137 ×7
8.	406 ×8	613 ×7	527 ×3	671 ×9	430 ×8
9.	941 ×7	483 ×8	675 ×9	729 ×8	436 ×7

CHAPTER 4

Lesson 5 Problem Solving

Solve each problem.

1. Each side of a baseball diamond is 90 feet in length. How far is it around the baseball diamond?

   It is _____ feet around the diamond.

2. Ashlee practices her flute 35 minutes a day. How many minutes does she practice during a week?

   She practices _____ minutes each week.

3. A certain light fixture contains six 75-watt lightbulbs. When all the bulbs are on, how many watts are being used?

   The six bulbs are using _____ watts.

4. A bus fare is 45 cents. What is the cost of two fares?

   The cost is _____ cents.

5. An aircraft carrier is as long as three football fields. A football field is 360 feet long. How long is the aircraft carrier?

   It is _____ feet long.

6. A shipment has five boxes. Each box weighs 125 kilograms. What is the weight of the shipment?

   The weight is _____ kilograms.

7. An airplane can travel 635 kilometers in one hour. How far can it travel in four hours?

   It can travel _____ kilometers in four hours.

8. A certain machine can produce 265 items an hour. At this rate, how many items can be produced in eight hours?

   _____ items can be produced.

1.	2.
3.	4.
5.	6.
7.	8.

CHAPTER 4 PRACTICE TEST
Multiplication (2- and 3-digit by 1-digit)

Multiply.

	a	*b*	*c*	*d*	*e*	*f*
1.	30 ×2	20 ×3	10 ×5	40 ×2	30 ×3	10 ×9
2.	32 ×3	43 ×2	21 ×4	23 ×3	11 ×6	12 ×4
3.	12 ×7	19 ×4	28 ×3	16 ×5	46 ×2	12 ×8
4.	60 ×7	80 ×9	40 ×7	71 ×6	92 ×4	81 ×6
5.	72 ×8	63 ×9	57 ×6	39 ×7	49 ×5	65 ×4

Multiply.

	a	*b*	*c*	*d*	*e*
6.	432 ×2	112 ×4	216 ×4	113 ×5	103 ×7
7.	131 ×7	282 ×3	711 ×5	612 ×4	932 ×3
8.	137 ×7	124 ×8	513 ×7	306 ×9	417 ×4
9.	521 ×9	941 ×8	567 ×4	439 ×5	857 ×6

Orange Book

CHAPTER 5 PRETEST
Multiplication (2- and 3-digit by 2-digit)

Multiply.

	a	*b*	*c*	*d*	*e*	*f*
1.	3 2 ×2 0	2 3 ×3 0	1 1 ×7 0	6 4 ×4 0	4 9 ×8 0	2 7 ×9 0
2.	4 3 ×2 1	2 3 ×3 2	3 4 ×2 2	2 3 ×3 9	2 1 ×4 8	3 2 ×3 7
3.	3 7 ×6 1	5 3 ×4 1	3 8 ×8 2	7 4 ×5 6	4 7 ×6 8	7 6 ×5 4

	a	*b*	*c*	*d*	*e*
4.	4 0 2 ×2 1	3 2 0 ×3 2	3 1 2 ×3 8	5 2 4 ×1 7	4 2 3 ×2 9
5.	6 2 4 ×6 1	2 1 3 ×5 3	4 3 1 ×8 2	4 2 6 ×5 7	8 3 4 ×6 8

Lesson 1 Multiplication (2-digit by 2-digit)

	Multiply 21 by 4 ones.	Multiply 21 by 3 tens.		
21 ×34	21 ×4 ―― 84	21 ×34 ―― 84 630	21 ×34 ―― 84 ⎫ 630 ⎬ Add. ―― 714	

Multiply.

	a	*b*	*c*	*d*	*e*	*f*
1.	1 3 ×3	1 3 ×3 0	4 3 ×2	4 3 ×2 0	4 2 ×2 0	2 3 ×3 0
2.	3 4 ×2	3 4 ×1 0	3 4 ×1 2	3 2 ×3	3 2 ×2 0	3 2 ×2 3
3.	2 1 ×4 2	3 3 ×3 2	7 9 ×1 1	3 1 ×3 1	1 1 ×2 7	1 2 ×3 2
4.	3 3 ×2 3	2 2 ×4 3	5 6 ×1 1	2 3 ×3 2	1 4 ×2 1	1 1 ×9 4

Lesson 1   Problem Solving

Solve each problem.

1. Each train car can carry ten vans. There are
   21 train cars full of vans. How many vans are
   there in all?

   Each train car can carry _____ vans.

   There are _____ train cars carrying vans.

   There are _____ vans in all.

2. Each train car can carry 15 automobiles. There are
   32 train cars full of automobiles. How many
   automobiles are there in all?

   Each train car can carry _____ automobiles.

   There are _____ train cars carrying
   automobiles.

   There are _____ automobiles in all.

3. Suppose there had been 48 train cars in
   Problem 2. Then, how many automobiles would
   there have been in all?

   There would have been _____ automobiles in all.

1.

2.

3.

Lesson 2　Multiplication (2-digit by 2-digit; renaming)

Multiply 47
by 2 ones.

Multiply 47
by 3 tens.

```
  47        47          47          47
 ×32       ×32         ×32         ×32
           ───         ───         ───
            94          94          94 ⎫ Add.
                      1410        1410 ⎭
                                  ─────
                                  1,504
```

Multiply.

	a	b	c	d	e	f
1.	65 ×7	65 ×70	37 ×8	37 ×80	64 ×40	75 ×50
2.	57 ×19	31 ×37	72 ×16	43 ×23	42 ×29	39 ×17
3.	83 ×91	43 ×62	44 ×52	32 ×63	34 ×32	52 ×41
4.	76 ×49	97 ×94	63 ×85	47 ×65	83 ×78	92 ×86

Lesson 2 Problem Solving

Solve each problem.

1. A school bus can carry 66 students. How many students can ride on 12 such buses?

   _____ students can ride on 12 buses.

2. A train can travel 97 kilometers in one hour. How far can it travel in 13 hours?

   It can travel _____ kilometers.

3. A copy machine can make 45 copies per minute. How many copies can it make in 15 minutes?

   It can make _____ copies in 15 minutes.

4. Ian bought 12 pizzas for a party. Each pizza was cut into 16 pieces. How many pieces did he have in all?

   He had _____ pieces in all.

5. Miss Lens bought 18 rolls of film. Thirty-six pictures can be taken on each roll. How many pictures can she take?

   _____ pictures can be taken.

6. Mr. Dzak works 40 hours each week. How many hours will he work in 26 weeks?

   He will work _____ hours.

7. A car is traveling 23 meters per second. How far will the car travel in 50 seconds?

   It will travel _____ meters.

8. There are 24 hours in a day. How many hours are there in 49 days?

   There are _____ hours in 49 days.

1.	2.
3.	4.
5.	6.
7.	8.

Lesson 3 Multiplication (2-digit by 2-digit; renaming)

Multiply.

	a	*b*	*c*	*d*	*e*	*f*
1.	2 4 ×2 0	3 2 ×3 0	2 1 ×4 0	5 6 ×2 0	7 5 ×6 0	8 4 ×7 0
2.	4 2 ×2 1	3 2 ×2 3	2 2 ×4 2	2 3 ×1 3	3 3 ×2 3	2 4 ×2 2
3.	4 3 ×2 5	8 7 ×1 7	3 4 ×2 5	3 2 ×2 8	2 1 ×4 8	3 2 ×3 9
4.	5 4 ×7 1	4 3 ×7 2	3 2 ×6 3	4 2 ×8 2	8 3 ×5 1	3 4 ×9 2
5.	6 8 ×7 3	4 2 ×5 8	4 9 ×8 6	3 7 ×9 4	6 2 ×4 8	2 8 ×5 9

CHAPTER 5

Lesson 3 Problem Solving

Solve each problem.

1. Mr. Carter can type 55 words a minute. How many words can he type in 15 minutes?

   He can type _____ words.

2. A machine puts caps on bottles at a rate of 96 per minute. At that rate, how many bottles can be capped in 25 minutes?

   _____ bottles can be capped in 25 minutes.

3. Mrs. Oliver travels 28 kilometers getting to and from work each day. How many kilometers will she travel in 22 working days?

   She will travel _____ kilometers.

4. There are 48 thumbtacks in a box. How many are there in 15 boxes?

   There are _____ thumbtacks.

5. There are 24 cars on a train. Suppose 66 passengers can ride in each car. How many passengers can ride in the train?

   _____ passengers can ride in the train.

6. Anthony delivers 75 papers each day. How many papers will he deliver in 14 days?

   He will deliver _____ papers.

7. A new building is to be 16 stories high. There are to be 14 feet for each story. How high will the building be?

   The building will be _____ feet high.

1.	
2.	3.
4.	5.
6.	7.

Lesson 4 Multiplication (3-digit by 2-digit; renaming)

Multiply 512
by 3 ones.

Multiply 512
by 2 tens.

```
  512          512             512            512
 ×23          ×23             ×23            ×23
             ─────           ─────          ─────
             1536            1536           1536 ⎫ Add.
                            10240          10240 ⎭
                                          ───────
                                           11,776
```

Multiply.

	a	b	c	d	e	f
1.	615 ×3	615 ×30	728 ×4	728 ×40	555 ×60	783 ×50
2.	132 ×4	132 ×20	132 ×24	323 ×3	323 ×60	323 ×63

Multiply.

	a	b	c	d	e
3.	212 ×23	423 ×12	121 ×49	321 ×37	412 ×24
4.	324 ×82	343 ×62	429 ×63	749 ×96	476 ×83

Lesson 4 Problem Solving

Solve each problem.

1. Chelsea plans to swim 150 laps this week. Each lap is 50 meters. How many meters does she plan to swim?

   She plans to swim _____ meters.

2. Each hour 225 pictures can be developed. How many pictures can be developed in 12 hours?

   _____ pictures can be developed.

3. One section of a sports arena has 24 rows of seats. There are 125 seats in each row. How many seats are there in that section?

   There are _____ seats in that section.

4. There are 144 bags of salt in a shipment. Each bag weighs 36 kilograms. What is the weight of the shipment?

   The weight is _____ kilograms.

5. James sells 165 papers a day. How many papers will he sell in 28 days?

   He will sell _____ papers.

6. A jet cruises at 575 miles an hour. At that rate, how many miles will it travel in 12 hours?

   It will travel _____ miles.

7. A certain desk weighs 118 kilograms. How many kilograms would 15 of the desks weigh?

   They would weigh _____ kilograms.

8. Twenty-three workers will deliver 290 circulars each. How many circulars will be delivered in all?

   _____ circulars will be delivered in all.

1.	2.
3.	4.
5.	6.
7.	8.

Lesson 5 Multiplication Review

Multiply.

	a	*b*	*c*	*d*	*e*	*f*
1.	54 ×21	75 ×42	63 ×39	27 ×64	84 ×56	67 ×93
2.	69 ×17	87 ×65	49 ×78	62 ×89	39 ×47	78 ×36

Multiply.

	a	*b*	*c*	*d*	*e*
3.	304 ×57	540 ×63	327 ×48	428 ×76	548 ×91
4.	924 ×27	286 ×14	478 ×82	375 ×89	721 ×52
5.	131 ×27	937 ×83	205 ×74	240 ×96	121 ×57

Lesson 5 Problem Solving

Solve each problem.

1. There are 24 slices of bread in a loaf. How many slices are there in 25 loaves?

   There are _____ slices.

2. There are 500 sheets in a giant pack of notebook paper. How many sheets are there in 12 giant packs?

   There are _____ sheets.

3. There are 12 eggs in a dozen. How many eggs are there in 16 dozen?

   There are _____ eggs in 16 dozen.

4. There are 180 eggs packed in a case. How many eggs are there in 24 cases?

   There are _____ eggs.

5. Fifty stamps are needed to fill each page of a stamp book. The book contains 24 pages. How many stamps are needed to fill the book?

   _____ stamps are needed.

6. Marsha practices the piano 35 minutes each day. How many minutes will she practice in 28 days?

   She will practice _____ minutes.

7. The average weight of the 11 starting players on a football team is 79 kilograms. What is the combined weight of the 11 players?

   _____ kilograms is the combined weight.

8. There are 328 pages in each of the 16 volumes of an encyclopedia. How many pages are there in all?

   There are _____ pages in all.

1.	2.
3.	4.
5.	6.
7.	8.

CHAPTER 5 PRACTICE TEST
Multiplication (2- and 3-digit by 2-digit)

Multiply.

	a	*b*	*c*	*d*	*e*
1.	21 ×43	23 ×13	34 ×21	34 ×25	23 ×36
2.	63 ×51	42 ×72	23 ×63	72 ×89	56 ×39
3.	314 ×21	123 ×32	302 ×23	527 ×15	607 ×41
4.	324 ×26	342 ×82	312 ×37	321 ×73	423 ×29
5.	682 ×57	428 ×96	357 ×85	537 ×49	729 ×84

Orange Book

CHAPTER 6 PRETEST
Multiplication (4-digit by 1- and 2-digit; 3-digit by 3-digit)

Multiply.

	a	*b*	*c*	*d*
1.	5 0 0 0 ×7	2 1 0 7 ×4	3 2 5 1 ×3	4 7 3 1 ×2
2.	7 1 3 1 ×5	7 6 5 2 ×8	2 1 2 1 ×2 4	6 7 4 2 ×1 7
3.	4 1 3 2 ×6 2	8 7 6 7 ×7 1	5 2 6 4 ×6 9	4 6 7 5 ×7 8
4.	3 2 1 ×3 0 0	5 6 7 ×4 0 0	3 1 2 ×3 2 0	4 3 2 ×2 0 7
5.	4 2 3 ×9 1 2	4 1 3 ×7 9 2	7 2 9 ×8 1 6	8 7 5 ×4 3 8

Lesson 1 Multiplication (4-digit by 1-digit)

		Multiply 2 ones by 4.	**Multiply 3 tens by 4.**	Multiply 2 hundreds by 4. Add the 1 hundred.	**Multiply 6 thousands by 4.**

$$\begin{array}{r} 6 \\ \times 4 \\ \hline 24 \end{array} \qquad \begin{array}{r} 6000 \\ \times 4 \\ \hline 24000 \end{array}$$

$$\begin{array}{r} 6232 \\ \times 4 \\ \hline 8 \end{array} \qquad \begin{array}{r} \overset{1}{6}232 \\ \times 4 \\ \hline 28 \end{array} \qquad \begin{array}{r} \overset{1}{6}232 \\ \times 4 \\ \hline 928 \end{array} \qquad \begin{array}{r} \overset{1}{6}232 \\ \times 4 \\ \hline 24928 \end{array}$$

Multiply.

	a	b	c	d	e	f
1.	$\begin{array}{r} 3 \\ \times 2 \\ \hline \end{array}$	$\begin{array}{r} 3000 \\ \times 2 \\ \hline \end{array}$	$\begin{array}{r} 2 \\ \times 4 \\ \hline \end{array}$	$\begin{array}{r} 2000 \\ \times 4 \\ \hline \end{array}$	$\begin{array}{r} 3 \\ \times 3 \\ \hline \end{array}$	$\begin{array}{r} 3000 \\ \times 3 \\ \hline \end{array}$
2.	$\begin{array}{r} 7 \\ \times 3 \\ \hline \end{array}$	$\begin{array}{r} 7000 \\ \times 3 \\ \hline \end{array}$	$\begin{array}{r} 8 \\ \times 5 \\ \hline \end{array}$	$\begin{array}{r} 8000 \\ \times 5 \\ \hline \end{array}$	$\begin{array}{r} 9 \\ \times 8 \\ \hline \end{array}$	$\begin{array}{r} 9000 \\ \times 8 \\ \hline \end{array}$

Multiply.

	a	b	c	d	e
3.	$\begin{array}{r} 3412 \\ \times 2 \\ \hline \end{array}$	$\begin{array}{r} 2018 \\ \times 4 \\ \hline \end{array}$	$\begin{array}{r} 1071 \\ \times 5 \\ \hline \end{array}$	$\begin{array}{r} 2731 \\ \times 3 \\ \hline \end{array}$	$\begin{array}{r} 8021 \\ \times 4 \\ \hline \end{array}$
4.	$\begin{array}{r} 1049 \\ \times 7 \\ \hline \end{array}$	$\begin{array}{r} 5107 \\ \times 8 \\ \hline \end{array}$	$\begin{array}{r} 1614 \\ \times 6 \\ \hline \end{array}$	$\begin{array}{r} 1751 \\ \times 5 \\ \hline \end{array}$	$\begin{array}{r} 5401 \\ \times 9 \\ \hline \end{array}$
5.	$\begin{array}{r} 5671 \\ \times 6 \\ \hline \end{array}$	$\begin{array}{r} 5407 \\ \times 9 \\ \hline \end{array}$	$\begin{array}{r} 4758 \\ \times 7 \\ \hline \end{array}$	$\begin{array}{r} 7034 \\ \times 5 \\ \hline \end{array}$	$\begin{array}{r} 6752 \\ \times 8 \\ \hline \end{array}$

Lesson 1 Problem Solving

Solve each problem.

1. Each tank truck can haul 5,000 gallons. How many gallons can be hauled by nine such trucks?

   _____ gallons can be hauled.

1.

2. Sample boxes of cereal were given out in seven cities. In each city 2,500 boxes were given out. How many boxes were given out in all?

   _____ boxes were given out.

2.

3. Eight autos are on a freight car. Each auto weighs 1,932 kilograms. What is the weight of all the autos on the freight car?

   The combined weight is _____ kilograms.

3.

4. The rail distance between St. Louis and San Francisco is 2,134 miles. How many miles does a train travel on a round trip between those cities?

   _____ miles are traveled.

4.

5. An airline attendant made five flights last week. The average length of each flight was 1,691 kilometers. How many kilometers did the attendant fly last week?

   The attendant flew _____ kilometers.

5.

6. There are nine machines in a warehouse. Each machine weighs 1,356 kilograms. What is the combined weight of the machines?

   The combined weight is _____ kilograms.

6.

7. The Gorman News Agency distributes 6,525 newspapers daily. How many newspapers does it distribute in six days?

   _____ newspapers are distributed in six days.

7.

Lesson 2 Multiplication (4-digit by 2-digit)

Multiply 5372
by 8 ones.

Multiply 5372
by 3 tens.

$$\begin{array}{r} 5372 \\ \times 38 \\ \hline 42976 \end{array}$$

$$\begin{array}{r} 5372 \\ \times 38 \\ \hline 42976 \\ 161160 \end{array}$$

$$\left.\begin{array}{r} 5372 \\ \times 38 \\ \hline 42976 \\ 161160 \\ \hline 204{,}136 \end{array}\right\} \text{Add.}$$

Multiply.

	a	*b*	*c*	*d*
1.	$\begin{array}{r} 4000 \\ \times 2 \end{array}$	$\begin{array}{r} 4000 \\ \times 20 \end{array}$	$\begin{array}{r} 5000 \\ \times 3 \end{array}$	$\begin{array}{r} 5000 \\ \times 30 \end{array}$
2.	$\begin{array}{r} 2000 \\ \times 40 \end{array}$	$\begin{array}{r} 3000 \\ \times 30 \end{array}$	$\begin{array}{r} 7000 \\ \times 50 \end{array}$	$\begin{array}{r} 6000 \\ \times 90 \end{array}$
3.	$\begin{array}{r} 2031 \\ \times 32 \end{array}$	$\begin{array}{r} 3132 \\ \times 22 \end{array}$	$\begin{array}{r} 2120 \\ \times 34 \end{array}$	$\begin{array}{r} 2314 \\ \times 25 \end{array}$
4.	$\begin{array}{r} 4312 \\ \times 28 \end{array}$	$\begin{array}{r} 8752 \\ \times 19 \end{array}$	$\begin{array}{r} 4321 \\ \times 72 \end{array}$	$\begin{array}{r} 3012 \\ \times 93 \end{array}$
5.	$\begin{array}{r} 7654 \\ \times 81 \end{array}$	$\begin{array}{r} 7542 \\ \times 65 \end{array}$	$\begin{array}{r} 8075 \\ \times 96 \end{array}$	$\begin{array}{r} 6209 \\ \times 58 \end{array}$

Lesson 2 Problem Solving

Solve each problem.

1. Each day 7,500 tons of ore can be processed. How many tons can be processed in 25 days?

   _____ tons can be processed.

2. A large ocean liner can carry 2,047 passengers. How many passengers can it carry on 36 trips?

   It can carry _____ passengers on 36 trips.

3. A computer can perform 9,456 computations per second. How many computations can it perform in 1 minute? (1 minute = 60 seconds)

   _____ computations can be performed.

4. Mr. LaFong earns $1,399 each month. How much will he earn in 12 months?

   He will earn $_____.

5. The Bulls played 82 basketball games last year. The attendance at each game was 6,547. What was the total attendance?

   The total attendance was _____.

6. There are 15 sections of reserved seats. Each section has 1,356 seats. How many reserved seats are there in all?

   There are _____ reserved seats.

7. Each day 2,228 cars can be assembled. How many cars can be assembled in 49 days?

   _____ cars can be assembled.

8. The Empire State Building is 1,472 feet tall. Mount Everest is 20 times taller than that. How tall is Mount Everest?

   Mount Everest is _____ feet tall.

1.	2.
3.	**4.**
5.	**6.**
7.	**8.**

Lesson 3 Multiplication (4-digit by 1- and 2-digit)

Multiply.

	a	*b*	*c*	*d*
1.	3 4 1 2 ×2	3 1 2 7 ×3	8 1 0 1 ×5	2 4 2 1 ×4
2.	1 3 0 7 ×6	8 1 7 2 ×4	9 7 0 1 ×9	7 0 6 1 ×8
3.	1 5 6 7 ×5	4 0 6 3 ×7	3 5 7 9 ×8	8 7 5 9 ×6
4.	2 3 0 1 ×2 3	4 7 5 3 ×7 1	4 3 2 1 ×8 2	3 0 1 2 ×9 3
5.	5 3 2 4 ×1 9	2 3 2 1 ×3 7	4 0 3 2 ×2 8	1 2 0 2 ×4 6
6.	5 6 7 8 ×5 7	9 3 0 5 ×7 8	6 0 7 8 ×9 6	8 3 4 9 ×7 4

Lesson 3   Problem Solving

Solve each problem.

1. Last week an average of 5,112 books a day were checked out of the city library. The library is open six days a week. How many books were checked out last week?

   _____ books were checked out.

2. Suppose books continue to be checked out at the rate indicated in problem **1**. How many books will be checked out in 26 days?

   _____ books will be checked out.

3. An auto dealer hopes to sell twice as many cars this year as last year. He sold 1,056 cars last year. How many cars does the dealer hope to sell this year?

   The dealer hopes to sell _____ cars.

4. The Humphreys drive an average of 1,245 miles each month. How many miles will they drive in a year? (1 year = 12 months)

   They will drive _____ miles.

5. The supermarket sells an average of 1,028 dozen eggs each week. How many dozen eggs will be sold in six weeks?

   _____ dozen eggs will be sold.

6. The Record Shoppe sells an average of 1,435 CDs each week. How many CDs will be sold in 52 weeks?

   _____ CDs will be sold.

7. A certain machine can produce 2,154 items an hour. How many items can be produced in eight hours?

   _____ items can be produced.

1.	
2.	**3.**
4.	**5.**
6.	**7.**

Lesson 4   Multiplication (3-digit by 3-digit)

Multiply 512
by 4 ones.

**Multiply 512
by 2 tens.**

Multiply 512
by 3 hundreds.

```
      512            512            512            512
     ×324           ×324           ×324           ×324
     ────           ────           ────           ────
     2048           2048           2048           2048  ⎫
                   10240          10240          10240  ⎬ Add.
                                 153600         153600  ⎭
                                               ─────────
                                               165,888
```

Multiply.

	a	b	c	d	e
1.	213 ×3	213 ×300	324 ×7	324 ×700	248 ×900
2.	321 ×213	223 ×239	342 ×260	213 ×823	423 ×257
3.	725 ×508	423 ×672	709 ×591	648 ×479	568 ×986

Lesson 4 Problem Solving

Flight 704 will have 125 passengers.

Solve each problem.

1. Suppose each passenger on Flight 704 has 100 pounds of luggage. What is the total weight of the luggage?

   The total weight is _____ pounds.

2. The average weight of each passenger is 135 pounds. What is the total weight of the passengers on Flight 704?

   The total weight is _____ pounds.

3. Each passenger on Flight 704 paid $103 for a ticket. How much money was paid in all?

   $_____ was paid in all.

4. The air distance from Chicago to Washington, D.C., is 957 kilometers. An airplane will make that flight 365 times this year. How many kilometers will be flown?

   _____ kilometers will be flown.

5. A jet was airborne 885 hours last year. Its average speed was 880 kilometers per hour. How far did the jet fly last year?

   The jet flew _____ kilometers last year.

1.

2.

3.

4.

5.

Lesson 5 Multiplication (3-digit by 3-digit)

Multiply.

	a	*b*	*c*	*d*
1.	203 ×213	143 ×121	432 ×128	213 ×216
2.	312 ×327	423 ×162	324 ×291	574 ×801
3.	234 ×712	212 ×934	567 ×148	423 ×278
4.	787 ×619	578 ×807	243 ×725	678 ×491
5.	785 ×580	584 ×787	375 ×998	673 ×578

CHAPTER 6

Lesson 5 Problem Solving

Solve each problem.

1. Last year a bookstore sold an average of 754 books on each of the 312 days it was open. How many books were sold last year?

   _____ books were sold.

2. In one hour, 560 loaves of bread can be baked. How many loaves can be baked in 112 hours?

   _____ loaves can be baked.

3. The garment factory can manufacture 960 shirts each day. How many shirts can be manufactured in 260 days?

   _____ shirts can be manufactured.

4. Approximately 925 rolls of newsprint are used each week in putting out the daily newspaper. How many rolls of newsprint will be needed in 104 weeks?

   _____ rolls of newsprint will be needed.

5. The average American uses 225 liters of water each day. How many liters of water does the average American use in 365 days?

   _____ liters of water are used in 365 days.

6. A factory has 109 shipping crates. Each crate can hold 148 pounds of goods. How many pounds of goods can be shipped in all the crates?

   _____ pounds of goods can be shipped.

7. A service station sells an average of 965 gallons of gasoline per day. How many gallons will be sold in 365 days?

   _____ gallons will be sold.

1.	
2.	**3.**
4.	**5.**
6.	**7.**

CHAPTER 6 PRACTICE TEST
Mulitiplication (4-digit by 1- and 2-digit; 3-digit by 3-digit)

Multiply.

	a	*b*	*c*	*d*
1.	2021 ×4	1107 ×6	4321 ×9	6758 ×8
2.	1022 ×23	2121 ×14	3241 ×27	3121 ×38
3.	5264 ×71	2134 ×82	5768 ×67	4938 ×89
4.	324 ×212	243 ×127	321 ×362	212 ×524
5.	584 ×167	778 ×518	432 ×692	789 ×694

CHAPTER 7 PRETEST
Money

For each dollar amount, write the digit that is in the given place value.

a	b	c	d
1. $10.30	$591.70	$324.90	$89.31
ones place	tens place	hundreds place	tenths place
_____	_____	_____	_____

Add or subtract.

a	b	c	d	e
2. 3 7¢ +4 2¢	$ 0.4 3 +0.6 4	1 7¢ 2 6¢ +4 2¢	$2.7 5 6.2 1 +3.6 2	$1 7.3 4 4 5.8 7 +2 5.1 2
3. 7 9¢ −2 7¢	$0.8 3 −0.2 4	$4.3 4 −0.7 2	$8.6 7 −4.8 9	$2 7.6 4 −1 8.6 9

Multiply.

4. $ 0.1 6 ×6	$ 0.4 6 ×7	$2.4 3 ×2	$1.1 7 ×1 9	$1 7.4 5 ×2 5

Solve each problem.

5. A suit costs $99.88. A sports coat costs $49.95. How much more does the suit cost than the sports coat?

The suit costs $_____ more.

6. You spent $1.98 in one store and $0.79 in another. How much did you spend in both stores?

You spent $_____ in both stores.

5.

6.

CHAPTER 7

Lesson 1 Decimal Place Value

In $1,265.84, what digit is in the tenths place? | Tell what place value the 4 is in for the dollar amount $26.34.

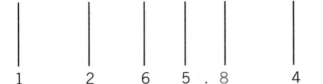

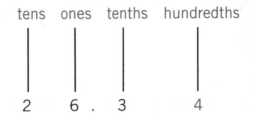

8 is in the tenths place. | 4 is in the hundredths place.

For each dollar amount, write the digit that is in the given place value.

	a	b	c	d
1.	$925.34	$2,683.95	$9,824.60	$5,871.64
	ones place	hundreds place	hundredths place	thousands place
	_____	_____	_____	_____
2.	$4,958.02	$7,196.28	$1,854.26	$6,295.08
	tens place	thousands place	tenths place	hundreds place
	_____	_____	_____	_____
3.	$956.42	$5,963.24	$9,147.23	$263.48
	tenths place	tens place	hundredths place	ones place
	_____	_____	_____	_____

Write what place value the given number is in for each exercise.

	a	b	c	d
4.	5 in $4,812.05	9 in $1,359.67	3 in $3,598.07	0 in $205.67
	_____	_____	_____	_____
5.	6 in $6,123.80	3 in $2,356.49	9 in $4,726.90	2 in $3,098.62
	_____	_____	_____	_____
6.	0 in $510.23	1 in $256.13	2 in $5,268.31	9 in $9,874.12
	_____	_____	_____	_____

Lesson 2 Writing Money

1 cent = 1¢ or $0.01
3 cents = ___3¢___ or ___$0.03___
65 cents = ___65¢___ or ___$0.65___

6 cents = ___6___ ¢ or $___0.06___

98 cents = _____¢ or $_____

1 dollar = $1.00
3 dollars and 2 cents = ___$3.02___
4 dollars and 59 cents = ___$4.59___

6 dollars and 3 cents = $___6.03___

5 dollars and 72 cents = $_____

Complete the following.

	a	b	c
1.	5 cents = _____¢	25¢ = $_____	$0.83 = _____¢
2.	10 cents = $_____	50¢ = $_____	$0.04 = _____¢
3.	25 cents = _____¢	75¢ = $_____	$0.29 = _____¢
4.	50 cents = $_____	10¢ = $_____	$0.06 = _____¢
5.	85 cents = _____¢	95¢ = $_____	$0.60 = _____¢
6.	100 cents = $_____	5¢ = $_____	$0.99 = _____¢

Complete the following.

	a	b
7.	4 dollars and 8 cents = $_____	$6.25 = 6 dollars and _____ cents
8.	7 dollars and 63 cents = $_____	$3.75 = _____ dollars and 75 cents
9.	3 dollars and 9 cents = $_____	$7.05 = 7 dollars and _____ cents
10.	6 dollars and 19 cents = $_____	$9.65 = _____ dollars and 65 cents
11.	5 dollars and 79 cents = $_____	$4.19 = _____ dollars and _____ cents
12.	18 dollars and 75 cents = $_____	$8.69 = _____ dollars and _____ cents

Lesson 3 Adding and Subtracting Money

Add the numbers.

25¢	$ 0.85
45¢	2.08
+19¢	+3.76
89¢	$6.69

Write ¢ or $ and a decimal point in the answer.

Subtract the numbers.

72¢	$12.07
−26¢	−4.83
46¢	$ 7.24

Write ¢ or $ and a decimal point in the answer.

Add or subtract.

	a	b	c	d	e
1.	2 3¢ +4 4¢ ¢	4 7¢ +2 5¢	$0.4 6 +0.7 3	$5.4 7 +8.2 1	$3 6.9 5 +7 2.0 2
2.	7 9¢ −2 3¢	5 6¢ −2 7¢	$1.2 7 −0.5 3	$4.6 7 −2.8 9	$3 6.7 8 −7.9 9
3.	1 4¢ +7 1¢	6¢ +8 7¢	$0.5 7 +0.6 8	$5.2 5 +9.4 6	$1 6.9 6 +2 7.4 5
4.	8 8¢ −6 9¢	9 2¢ −8 9¢	$2.6 4 −0.5 7	$6.2 7 −2.8 9	$4 9.7 8 −1 8.8 9
5.	1 2¢ 3 9¢ +2 4¢	4 3¢ 2 7¢ +2 6¢	$0.7 5 0.6 5 +0.9 7	$0.1 2 4.6 9 +5.8 7	$4 7.5 2 8 9.2 5 +6 7.4 7
6.	$2.4 6 −0.8 7	$1.5 7 −0.9 9	$3.0 7 −1.8 5	$7.0 0 −2.4 8	$6 0.4 7 −2 7.5 9

Lesson 3 Problem Solving

Solve each problem.

1. Elisa had a $2.50 hot dog and a $0.67 milk for lunch. How much did she spend on her lunch?

   Elisa spent _____ on her lunch.

2. Branford paid $2.25 for a notebook and $1.79 for new pens. How much did he pay in all?

   Branford spent _____ in all.

3. Cherise bought a movie ticket for $5.75. She gave the clerk $10.00. How much change should she receive?

   Cherise should receive _____ in change.

4. Samuel bought a DVD for $14.79. He gave the clerk a twenty-dollar bill. How much change will he get back?

   Samuel will get back _____ .

5. Bert had a bagel sandwich that cost $2.69 and an order of cottage cheese that cost $0.75. How much did he spend?

   Bert spent _____ .

6. Emily bought a model car for $5.24. She gave the clerk $10.29. How much change did the clerk give Emily?

   Emily got _____ in change.

7. Manuela bought a computer game for $13.50. She gave the clerk $15.00. How much change did the clerk give Manuela?

   The clerk gave Manuela _____ .

1.
2.
3.
4.
5.
6.
7.

Lesson 4   Multiplying Money

$$
\begin{array}{r}
\$0.19 \\
\times 4 \\
\hline
\$0.76
\end{array}
$$

Multiply the numbers. Write $ and a decimal point in the answer.

Make sure there are two digits to the right of the decimal point.

$$
\begin{array}{r}
\$3.24 \\
\times 46 \\
\hline
1944 \\
12960 \\
\hline
\$149.04
\end{array}
$$

Multiply.

	a	b	c	d	e
1.	$0.2 1 ×4	$0.1 3 ×5	$0.2 6 ×3	$1.7 2 ×4	$1 6.2 3 ×3
2.	$0.3 1 ×6	$0.5 2 ×7	$0.5 8 ×6	$2.4 7 ×4	$2 5.7 9 ×3
3.	$0.3 2 ×1 2	$0.4 3 ×2 3	$0.3 4 ×5 2	$4.4 9 ×3 6	$4 3.7 5 ×2 4
4.	$2.3 8 ×4 2 6	$3.0 9 ×3 2 9	$1.2 4 ×1 0 2	$4.2 9 ×3 1 7	$7 5.9 6 ×4 4

CHAPTER 7

Lesson 4   Problem Solving

Solve each problem.

1. Find the cost of these four items: soap, $0.58; soup, $0.55; napkins, $0.87; and toothpicks, $0.39.

   The total cost is $_____.

2. A quart of motor oil sells for $1.49 at a service station. It costs $0.99 at a discount store. What is the difference between those prices?

   The difference is _____ ¢.

3. Spark plugs are on sale at $2.19 each. What is the cost of eight spark plugs?

   The cost is $_____.

4. It costs $19.75 to go to River City by train. It costs $16.95 to go by bus. How much cheaper is it to go by bus?

   It is $_____ cheaper.

5. Tires cost $89.89 each. What is the cost of four tires?

   The cost is $_____.

6. A ticket for a bleacher seat costs $3.75. A ticket for a box seat costs $9.50. How much more does it cost to sit in a box seat than in a bleacher seat?

   It costs $_____ more.

7. How much money have the four students named in the table saved?

   They have saved $_____.

8. How much more money has Cary saved than Jerri?

   Cary has saved $_____ more.

Student	Amount saved
Barry	$27.49
Cary	$33.14
Jerri	$29.36
Mary	$28.76

1.	**2.**
3.	**4.**
5.	**6.**
7.	**8.**

CHAPTER 7 PRACTICE TEST
Money

For each dollar amount, write the digit that is in the given place value.

a	b	c	d
1. $3,295.12	$543.06	$2,536.70	$295.14
tenths place	hundreds place	hundredths place	ones place
_____	_____	_____	_____

Write the place value for the given number.

a	b	c	d
2. 5 in $2,561.03	9 in $156.92	3 in $5,164.93	0 in $908.67
_____	_____	_____	_____

Add.

	a	b	c	d
3.	52¢ +13¢	76¢ +14¢	$8.25 + 3.78	$41.53 27.15 +12.36

Subtract.

4.	75¢ − 23¢	67¢ −29¢	$7.62 − 3.49	$10.00 − 6.67

Multiply.

5.	$0.38 × 4	$2.25 × 6	$5.29 × 12	$4.24 × 21

6. Thomas bought a booster pack for $2.98, some football cards for $5.67, and a super-sized candy bar for $1.29. How much did he spend in all?

Thomas spent a total of $_____ .

6.

7. A front row ticket at the concert costs $68.50. A ticket for the bleachers costs $42.95. How much more does it cost to sit in the front row than in the bleachers?

A front row ticket costs $_____ more than a bleacher ticket.

7.

CHAPTER 7

CHAPTER 8 PRETEST
Division (basic facts through 81 ÷ 9)

Divide.

	a	*b*	*c*	*d*	*e*
1.	$5\overline{)5}$	$1\overline{)4}$	$3\overline{)12}$	$2\overline{)4}$	$4\overline{)28}$
2.	$1\overline{)0}$	$2\overline{)18}$	$4\overline{)8}$	$3\overline{)24}$	$5\overline{)35}$
3.	$6\overline{)24}$	$6\overline{)12}$	$6\overline{)48}$	$6\overline{)6}$	$6\overline{)30}$
4.	$6\overline{)42}$	$6\overline{)18}$	$6\overline{)36}$	$6\overline{)54}$	$6\overline{)0}$
5.	$7\overline{)56}$	$7\overline{)42}$	$7\overline{)0}$	$7\overline{)28}$	$7\overline{)14}$
6.	$7\overline{)63}$	$7\overline{)21}$	$7\overline{)49}$	$7\overline{)7}$	$7\overline{)35}$
7.	$8\overline{)56}$	$8\overline{)40}$	$8\overline{)24}$	$8\overline{)72}$	$8\overline{)8}$
8.	$8\overline{)16}$	$8\overline{)64}$	$8\overline{)0}$	$8\overline{)32}$	$8\overline{)48}$
9.	$9\overline{)54}$	$9\overline{)9}$	$9\overline{)72}$	$9\overline{)36}$	$9\overline{)18}$
10.	$9\overline{)81}$	$9\overline{)63}$	$9\overline{)27}$	$9\overline{)0}$	$9\overline{)45}$

Lesson 1   Division (facts through 45 ÷ 5)

$$\overset{4}{3\overline{)12}}$$ → Find the 12 in
 → the **3** -column.

The quotient is named
in the ▨ at the end
of this row.

Use the table to divide.

$$5\overline{)30}$$

×	0	1	2	3	4	5	6	7	8	9
0	0	0	0	0	0	0	0	0	0	0
1	0	1	2	3	4	5	6	7	8	9
2	0	2	4	6	8	10	12	14	16	18
3	0	3	6	9	12	15	18	21	24	27
4	0	4	8	12	16	20	24	28	32	36
5	0	5	10	15	20	25	30	35	40	45
6	0	6	12	18	24	30	36	42	48	54
7	0	7	14	21	28	35	42	49	56	63
8	0	8	16	24	32	40	48	56	64	72
9	0	9	18	27	36	45	54	63	72	81

Divide.

	a	b	c	d	e	f
1.	$3\overline{)15}$	$4\overline{)12}$	$1\overline{)7}$	$5\overline{)25}$	$2\overline{)12}$	$4\overline{)28}$
2.	$5\overline{)35}$	$2\overline{)14}$	$3\overline{)18}$	$4\overline{)20}$	$1\overline{)9}$	$3\overline{)9}$
3.	$4\overline{)32}$	$5\overline{)20}$	$1\overline{)6}$	$3\overline{)12}$	$2\overline{)10}$	$3\overline{)21}$
4.	$2\overline{)16}$	$3\overline{)3}$	$4\overline{)16}$	$1\overline{)8}$	$4\overline{)0}$	$5\overline{)10}$
5.	$3\overline{)24}$	$5\overline{)0}$	$2\overline{)8}$	$4\overline{)36}$	$5\overline{)15}$	$3\overline{)27}$
6.	$4\overline{)24}$	$2\overline{)18}$	$5\overline{)40}$	$3\overline{)0}$	$1\overline{)5}$	$5\overline{)45}$
7.	$1\overline{)4}$	$2\overline{)6}$	$5\overline{)30}$	$5\overline{)5}$	$4\overline{)8}$	$3\overline{)6}$

CHAPTER 8

Lesson 1 Problem Solving

Solve each problem.

1. There are 24 cars in the parking lot. There are
   three rows of cars. Each row has the same number
   of cars. How many cars are in each row?

   There are _____ cars.

   There are _____ rows.

   There are _____ cars in each row.

1.

2. Five people went to the zoo. The tickets cost $25 in
   all. Each ticket costs the same amount. How much
   did each ticket cost?

   Each ticket costs $_____.

3. There are 16 people playing ball. There are two
   teams. Each team has the same number of players.
   How many people are on each team?

   _____ people are on each team.

2.   **3.**

4. Four people are seated at each table. There are
   16 people in all. How many tables are there?

   There are _____ tables.

5. Two people are in each boat. There are 16 people in
   all. How many boats are there?

   There are _____ boats.

4.   **5.**

Lesson 2 Division (facts through 63 ÷ 7)

$$5 \dashrightarrow 5$$
$$\underline{\times 6} \dashrightarrow 6\overline{)\,30}$$
$$30 \dashrightarrow$$

$$5 \dashrightarrow 5$$
$$\underline{\times 7} \dashrightarrow 7\overline{)\,35}$$
$$35 \dashrightarrow$$

$6 \times 5 = 30$, so $30 \div 6 = \underline{\ 5\ }$.

$7 \times 5 = 35$, so $35 \div 7 = \underline{\qquad}$.

Complete the following.

	a	b	c	d
1.	6 $\underline{\times 6}$ so $6\overline{)\,36}$ 36	7 $\underline{\times 6}$ so $6\overline{)\,42}$ 42	8 $\underline{\times 6}$ so $6\overline{)\,48}$ 48	9 $\underline{\times 6}$ so $6\overline{)\,54}$ 54
2.	6 $\underline{\times 7}$ so $7\overline{)\,42}$ 42	7 $\underline{\times 7}$ so $7\overline{)\,49}$ 49	8 $\underline{\times 7}$ so $7\overline{)\,56}$ 56	9 $\underline{\times 7}$ so $7\overline{)\,63}$ 63

Divide.

	a	b	c	d	e
3.	$6\overline{)\,54}$	$7\overline{)\,14}$	$6\overline{)\,30}$	$7\overline{)\,35}$	$6\overline{)\,6}$
4.	$7\overline{)\,49}$	$6\overline{)\,42}$	$7\overline{)\,42}$	$6\overline{)\,0}$	$7\overline{)\,21}$
5.	$6\overline{)\,18}$	$7\overline{)\,7}$	$6\overline{)\,36}$	$7\overline{)\,56}$	$6\overline{)\,24}$
6.	$7\overline{)\,28}$	$6\overline{)\,48}$	$7\overline{)\,0}$	$6\overline{)\,12}$	$7\overline{)\,63}$
7.	$4\overline{)\,16}$	$4\overline{)\,28}$	$5\overline{)\,20}$	$5\overline{)\,35}$	$4\overline{)\,36}$

CHAPTER 8

Lesson 2 Problem Solving

Solve each problem.

1. Mrs. Shields had 24 plants. She put them into rows of 6 plants each. How many rows were there?

   Mrs. Shields had _____ plants.

   She put them into rows of _____ plants each.

   She had _____ rows of plants.

2. Twenty-eight students sat at seven tables. The same number of students sat at each table. How many students sat at each table?

   There were _____ students.

   There were _____ tables.

   _____ students sat at each table.

3. Pencils cost 7¢ each. Kelley has 63¢. How many pencils can she buy?

   She can buy _____ pencils.

4. Mr. Rojas put 56 books into stacks of 7 books each. How many stacks of books did he have?

   He had _____ stacks of books.

5. Marion Street has 54 streetlights. Each block has 6 lights. How many blocks are there?

   There are _____ blocks.

6. There are 35 days before Alex's birthday. How many weeks is it before his birthday?
   (7 days = 1 week)

   There are _____ weeks before Alex's birthday.

7. There are 48 items to be packed. Six items can be packed in each box. How many boxes are needed?

   _____ boxes are needed.

1.	
2.	
3.	4.
5.	6.
7.	

Lesson 3   Division (facts through 81 ÷ 9)

```
 4  - - - - - - →    4
×8  - - - - →    8) 3 2
───                    ↑
 32  - - - - - - - - - ┘
```

```
 5  - - - - - - →    5
×9  - - - - - →    9) 4 5
───                    ↑
 45  - - - - - - - - - ┘
```

8 × 4 = 32, so 32 ÷ 8 = __4__ .

9 × 5 = 45, so 45 ÷ 9 = _____ .

Complete the following.

	a	b	c	d
1.	6 ×8 so 8) 48 ─ 48	7 ×8 so 8) 56 ─ 56	8 ×8 so 8) 64 ─ 64	9 ×8 so 8) 72 ─ 72
2.	6 ×9 so 9) 54 ─ 54	7 ×9 so 9) 63 ─ 63	8 ×9 so 9) 72 ─ 72	9 ×9 so 9) 81 ─ 81

Divide.

	a	b	c	d	e
3.	8) 8	9) 1 8	8) 5 6	9) 2 7	8) 3 2
4.	9) 4 5	8) 6 4	9) 0	8) 2 4	9) 5 4
5.	8) 7 2	9) 3 6	8) 1 6	9) 9	8) 4 0
6.	9) 6 3	8) 0	9) 7 2	8) 4 8	9) 8 1
7.	7) 4 2	6) 4 8	7) 5 6	6) 5 4	7) 6 3

Lesson 3 Problem Solving

Solve each problem.

1. A checkerboard has 64 squares. Each of the eight rows has the same number of squares. How many squares are in each row?

   _____ squares are in each row.

2. Thirty-six children came to the park to play ball. How many teams of nine players each could be formed?

   _____ teams could be formed.

3. Each washer load weighs nine pounds. How many washer loads are there in 54 pounds of laundry?

   _____ loads are in 54 pounds of laundry.

4. There are 48 apartments on eight floors. There is the same number of apartments on each floor. How many apartments are there on each floor?

   _____ apartments are on each floor.

5. Nine books weigh 18 pounds. Each book has the same weight. How much does each book weigh?

   Each book weighs _____ pounds.

6. Miss McKee can input 8 pages an hour on the computer. How long would it take her to input 24 pages?

   It would take her _____ hours.

7. How many tosses can you make for 72¢?

   _____ tosses can be made.

8. Andrea tossed seven bags through the same hole. Her score was 49. How many points did she score on each toss?

   She scored _____ points on each toss.

Beanbag Toss
8¢ per toss

6

7   8

1.	2.
3.	4.
5.	6.
7.	8.

Lesson 4 Division (facts through 81 ÷ 9)

Complete the following.

	a	*b*	*c*	*d*

1.
$$\begin{array}{r} 4 \\ \times 9 \\ \hline 36 \end{array} \text{ so } 9\overline{)36}$$
$$\begin{array}{r} 7 \\ \times 2 \\ \hline 14 \end{array} \text{ so } 2\overline{)14}$$
$$\begin{array}{r} 8 \\ \times 1 \\ \hline 8 \end{array} \text{ so } 1\overline{)8}$$
$$\begin{array}{r} 7 \\ \times 6 \\ \hline 42 \end{array} \text{ so } 6\overline{)42}$$

2.
$$\begin{array}{r} 6 \\ \times 7 \\ \hline 42 \end{array} \text{ so } 7\overline{)42}$$
$$\begin{array}{r} 8 \\ \times 3 \\ \hline 24 \end{array} \text{ so } 3\overline{)24}$$
$$\begin{array}{r} 6 \\ \times 4 \\ \hline 24 \end{array} \text{ so } 4\overline{)24}$$
$$\begin{array}{r} 0 \\ \times 5 \\ \hline 0 \end{array} \text{ so } 5\overline{)0}$$

Divide.

	a	*b*	*c*	*d*	*e*
3.	$4\overline{)36}$	$9\overline{)72}$	$6\overline{)54}$	$5\overline{)10}$	$8\overline{)56}$
4.	$2\overline{)6}$	$4\overline{)20}$	$7\overline{)28}$	$1\overline{)4}$	$6\overline{)30}$
5.	$7\overline{)56}$	$6\overline{)18}$	$3\overline{)0}$	$9\overline{)54}$	$8\overline{)40}$
6.	$1\overline{)6}$	$9\overline{)18}$	$5\overline{)20}$	$7\overline{)14}$	$4\overline{)12}$
7.	$2\overline{)2}$	$2\overline{)18}$	$8\overline{)24}$	$6\overline{)6}$	$1\overline{)2}$
8.	$7\overline{)0}$	$3\overline{)12}$	$3\overline{)6}$	$9\overline{)0}$	$5\overline{)30}$
9.	$5\overline{)40}$	$8\overline{)8}$	$2\overline{)10}$	$1\overline{)0}$	$4\overline{)4}$
10.	$3\overline{)18}$	$7\overline{)35}$	$8\overline{)32}$	$9\overline{)27}$	$2\overline{)16}$

CHAPTER 8

Lesson 4 Problem Solving

Solve each problem.

1. There are 45 school days left before vacation. There are 5 school days each week. How many weeks are left before vacation?

   There are _____ weeks left.

2. There are 35 seats in Mrs. Champney's room. The seats are arranged in seven rows with the same number in each row. How many seats are there in each row?

   There are _____ seats in each row.

3. Diane bought 21 feet of material. How many yards of material did she buy? (3 feet = 1 yard)

   Diane bought _____ yards of material.

4. Forty-eight cars are parked in a parking lot. The cars are parked in six rows with the same number in each row. How many cars are parked in each row?

   _____ cars are parked in each row.

5. There are 36 seats on a bus and 4 seats per row. How many rows of seats are there on the bus?

   There are _____ rows of seats on the bus.

6. Mr. Woods works a six-day week. He works 54 hours each week and the same number of hours each day. How many hours does he work each day?

   He works _____ hours each day.

7. It takes 9 minutes to assemble a certain item. How many items can be assembled in 54 minutes?

   _____ items can be assembled.

8. There are 32 students in a relay race. Four run on each team. How many teams are there?

   There are _____ teams.

1.	2.
3.	4.
5.	6.
7.	8.

CHAPTER 8
Division (basic facts through 81 ÷ 9)

112

Lesson 4
Division (facts through 81 ÷ 9)

CHAPTER 8 PRACTICE TEST
Division (basic facts through 81 ÷ 9)

Divide.

	a	b	c	d	e
1.	8)‾1‾6	5)‾1‾5	2)‾0	6)‾4‾8	9)‾8‾1
2.	4)‾8	1)‾1	3)‾3	8)‾7‾2	7)‾6‾3
3.	3)‾1‾8	4)‾3‾2	7)‾4‾2	2)‾4	5)‾5
4.	7)‾4‾9	5)‾2‾5	6)‾3‾6	9)‾9	2)‾8
5.	6)‾4‾2	2)‾1‾2	8)‾0	1)‾3	4)‾2‾8
6.	3)‾1‾5	6)‾2‾4	1)‾5	8)‾3‾2	4)‾2‾4
7.	5)‾3‾5	1)‾7	6)‾1‾2	7)‾7	9)‾2‾7
8.	3)‾9	4)‾1‾6	2)‾1‾6	3)‾2‾1	1)‾9
9.	8)‾4‾8	7)‾2‾1	5)‾4‾5	9)‾4‾5	6)‾0
10.	9)‾6‾3	4)‾0	3)‾2‾7	7)‾3‾5	8)‾6‾4

CHAPTER 8

CHAPTER 9 PRETEST
Division (2- and 3-digit by 1-digit)

Divide.

	a	*b*	*c*	*d*	*e*
1.	3)26	5)47	6)49	8)62	9)77
2.	2)28	4)48	3)93	7)84	6)96
3.	5)67	8)93	9)97	6)87	7)97
4.	3)186	4)236	7)161	9)425	8)612
5.	3)369	4)840	2)964	5)696	7)898

Lesson 1 Division (2-digit)

Study how to divide 32 by 6.

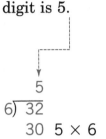

×	1	2	3	4	5	6
6	6	12	18	24	30	36

32 is between 30 and 36, so 32÷6 is between 5 and 6. The ones digit is 5.

Since 32−30=2 and 2 is less than 6, the **remainder** 2 is recorded like this:

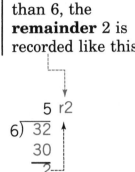

```
     5
6) 32
   30   5 × 6
    2   Subtract.
```

```
     5 r2
6) 32
   30
    2
```

Study how to divide 43 by 9.

×	1	2	3	4	5	6
9	9	18	27	36	45	54

43 is between 36 and 45, so 43÷9 is between 4 and 5. The ones digit is 4.

Since 43−36=7 and 7 is less than 9, the **remainder** is 7.

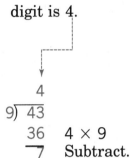

```
     4
9) 43
   36   4 × 9
    7   Subtract.
```

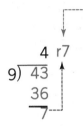

```
     4 r7
9) 43
   36
    7
```

Divide.

	a	b	c	d	e
1.	5) 2 7	8) 4 7	2) 1 7	9) 4 6	6) 3 8
2.	7) 6 1	9) 6 7	5) 4 9	3) 2 3	8) 7 8
3.	4) 3 8	6) 4 5	8) 6 3	2) 1 9	7) 3 8

Lesson 1　Problem Solving

Solve each problem.

1. How many pieces of rope 4 meters long can be cut from 15 meters of rope? How much rope will be left over?

   _____ pieces can be cut.

   _____ meters will be left over.

2. Dale has 51¢. Magic rings cost 8¢ each. How many magic rings can he buy? How much money will he have left?

   Dale can buy _____ magic rings.

   He will have _____ ¢ left.

3. Maria has a board that is 68 inches long. How many 9-inch pieces can be cut from this board? How long is the piece that is left?

   _____ pieces can be cut.

   _____ inches will be left.

4. Twenty-eight liters of water was used to fill buckets that hold 5 liters each. How many buckets were filled? How much water was left over?

   _____ buckets were filled.

   _____ liters of water were left over.

5. It takes 8 minutes to make a doodad. How many doodads can be made in 60 minutes? How much time would be left to begin making another doodad?

   _____ doodads can be made.

   _____ minutes would be left.

1.

2.

3.

4.

5.

Lesson 2   Division (2-digit)

Study how to divide 72 by 3.

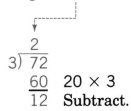

×	10	20	30	40
3	30	60	90	120

72 is between 60 and 90, so
72÷3 is between 20 and 30.
The tens digit is 2.

$$\begin{array}{r} 2 \\ 3\overline{)72} \\ \underline{60} \quad 20 \times 3 \\ 12 \quad \text{Subtract.} \end{array}$$

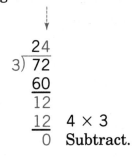

×	1	2	3	4	5
3	3	6	9	12	15

4×3=12, so the ones
digit is 4.

$$\begin{array}{r} 24 \\ 3\overline{)72} \\ \underline{60} \\ 12 \\ \underline{12} \quad 4 \times 3 \\ 0 \quad \text{Subtract.} \end{array}$$

CHAPTER 9

Divide.

	a	b	c	d	e
1.	2)26	4)48	5)55	3)96	4)88
2.	4)47	6)69	3)65	7)96	8)99
3.	7)91	9)96	3)87	8)97	4)92

Lesson 2 Problem Solving

Solve each problem.

1. Ninety-nine students are to be separated into nine groups. The same number of students is to be in each group. How many students will be in each group?

   _____ students will be in each group.

2. There are 94 grapefruit in a crate. How many bags of 6 grapefruit each can be filled by using the grapefruit from one crate? How many grapefruit will be left over?

   _____ bags can be filled.

   _____ grapefruit will be left over.

3. Ms. McClean's lot is 60 feet wide. What is the width of the lot in yards? (3 feet = 1 yard)

   Her lot is _____ yards wide.

4. There are 96 fluid ounces of punch in a bowl. How many 7-ounce glasses can be filled? How many fluid ounces will be left over?

   _____ glasses can be filled.

   _____ fluid ounces will be left over.

5. Toy cars are packed into boxes of 8 cars each. How many boxes will be needed to pack 96 toy cars?

   _____ boxes will be needed.

6. You have 64 pennies to exchange for nickels. How many nickels will you get? How many pennies will be left?

   You will get _____ nickels.

   _____ pennies will be left.

1.

2.

3.

4.

5.

6.

Lesson 3 Division (2-digit)

Divide.

	a	*b*	*c*	*d*	*e*
1.	3)2̄3̄	5)4̄6̄	7)6̄7̄	9)8̄5̄	8)6̄9̄
2.	2)2̄8̄	3)3̄6̄	4)4̄4̄	3)9̄3̄	4)8̄4̄
3.	5)8̄5̄	6)9̄6̄	8)9̄6̄	7)9̄8̄	9)9̄0̄
4.	7)7̄4̄	5)5̄7̄	8)9̄9̄	2)8̄5̄	3)6̄5̄
5.	3)7̄9̄	5)6̄7̄	6)9̄7̄	4)9̄4̄	7)8̄9̄

Lesson 3 Problem Solving

Solve each problem.

1. Paul has 75 stamps. His album will hold 9 stamps per page. How many pages can he fill? How many stamps will be left over?

   He can fill _____ pages.

   _____ stamps will be left over.

2. Kelsey has 96 stamps. Her album will hold 6 stamps per page. How many pages can she fill?

   She can fill _____ pages.

3. Seventy-seven stamps are given to a stamp club. Each of the seven members is to receive the same number of stamps. How many stamps will each member get?

   Each of the members will get _____ stamps.

4. Anne has 87 cents. What is the greatest number of nickels she could have? What is the least number of pennies she could have?

   She could have at most _____ nickels.

   She could have as few as _____ pennies.

5. Fifty-four students are to be separated into teams of nine students each. How many teams can be formed?

   _____ teams can be formed.

6. Rex has 63 empty bottles. He wants to put them into cartons of 8 bottles each. How many cartons can he fill? How many bottles will be left over?

   _____ cartons can be filled.

   _____ bottles will be left over.

1.	2.
3.	4.
5.	6.

Lesson 4   Division (3-digit)

Study how to divide 263 by 5.

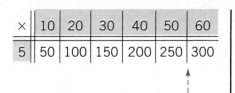

×	10	20	30	40	50	60
5	50	100	150	200	250	300

×	1	2	3	4	5
5	5	10	15	20	25

Since 100 × 5 = 500 and 500 is greater than 263, there is no hundreds digit.

$5\overline{)263}$

263 is between 250 and 300.
263 ÷ 5 is between 50 and 60.
The tens digit is 5.

```
      5
5) 263
   250    50 × 5
    13    Subtract.
```

13 is between 10 and 15.
13 ÷ 5 is between 2 and 3.
The ones digit is 2.

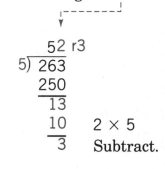

```
     52 r3
5) 263
   250
    13
    10    2 × 5
     3    Subtract.
```

CHAPTER 9

Divide.

	a	b	c	d	e
1.	$4\overline{)248}$	$6\overline{)366}$	$3\overline{)189}$	$7\overline{)266}$	$8\overline{)472}$
2.	$9\overline{)547}$	$2\overline{)121}$	$5\overline{)308}$	$6\overline{)374}$	$4\overline{)341}$
3.	$8\overline{)735}$	$3\overline{)252}$	$9\overline{)479}$	$7\overline{)378}$	$5\overline{)473}$

Lesson 4 Problem Solving

Solve each problem.

1. A truck driver drove from Chicago to St. Louis in six hours. The same distance was driven each hour. How many kilometers were driven each hour?

_____ kilometers were driven each hour.

2. A bus left New York City and arrived in Baltimore four hours later. The bus went the same distance each hour. How many kilometers were traveled each hour?

_____ kilometers were traveled each hour.

3. Suppose the same distance is driven each hour. How many kilometers must be driven each hour to go from Los Angeles to San Francisco in eight hours?

_____ kilometers must be driven each hour.

4. The Jetts' plan is to drive from Los Angeles to Phoenix in seven hours. Suppose they travel the same distance each hour. How many kilometers must they travel each hour?

_____ kilometers must be driven each hour.

5. Suppose the same distance is traveled each hour. How many kilometers must be traveled each hour to go from Chicago to Pittsburgh in nine hours?

_____ kilometers must be traveled each hour.

1.	
2.	**3.**
4.	**5.**

Lesson 5   Division (3-digit)

Study how to divide 813 by 4.

×	100	200	300
4	400	800	1200

813 is between 800 and 1200, so 813÷4 is between 200 and 300. The hundreds digit is 2.

```
    2
4) 813
   800   200 × 4
   ‾‾‾
    13   Subtract.
```

Since 10 × 4 = 40 and 40 is greater than 13, the tens digit is 0.

```
   20
4) 813
   800
   ‾‾‾
    13
     0   0 × 4
   ‾‾‾
    13   Subtract.
```

×	1	2	3	4
4	4	8	12	16

13 is between 12 and 16, so 13÷4 is between 3 and 4. The ones digit is 3.

```
   203 r1
4) 813
   800
   ‾‾‾
    13
     0
   ‾‾‾
    13
    12   3 × 4
   ‾‾
     1   Subtract.
```

CHAPTER 9

Divide.

	a	b	c	d	e
1.	2) 4 6 8	4) 4 7 2	3) 6 0 9	5) 5 8 5	7) 8 8 2
2.	8) 8 7 6	6) 7 9 4	9) 9 7 9	2) 9 8 7	5) 5 9 3
3.	6) 8 4 2	3) 9 4 9	7) 8 7 5	4) 8 7 9	8) 9 9 2

Lesson 5 Problem Solving

Solve each problem.

1. Mrs. Steel needs 960 trading stamps to fill a book. These stamps will fill eight pages with the same number of stamps on each page. How many stamps are needed to fill each page?

   _____ stamps are needed to fill each page.

2. There are 576 pencils in four cases. There are the same number of pencils in each case. How many pencils are in each case?

   There are _____ pencils in each case.

3. There are 532 apples in all. How many sacks of 5 apples each can be filled? How many apples will be left over?

   _____ sacks can be filled.

   _____ apples will be left over.

4. At a factory, 968 items were manufactured during an eight-hour shift. The same number was manufactured each hour. How many items were manufactured each hour?

   _____ items were manufactured each hour.

5. The Chesapeake Bridge is 540 feet long. It consists of four sections of the same length. How long is each section?

   Each section is _____ feet long.

6. A full load for a dry cleaning machine is 5 suits. There are 624 suits to be cleaned. How many full loads will there be? How many suits will be in the partial load?

   There will be _____ full loads.

   There will be _____ suits in the partial load.

7. A carpenter uses 5 nails to shingle one square foot of roof. At that rate, how many square feet of roof could be shingled by using 750 nails?

   _____ square feet could be shingled.

1.	
2.	**3.**
4.	**5.**
6.	**7.**

Lesson 6 Division (3-digit)

Divide.

	a	*b*	*c*	*d*	*e*
1.	2)‾1 2 6	6)‾4 8 6	3)‾2 4 9	7)‾5 5 3	8)‾6 2 4
2.	5)‾4 7 3	4)‾3 5 7	9)‾7 5 8	6)‾5 2 5	3)‾2 6 9
3.	3)‾6 9 3	2)‾8 1 6	4)‾8 5 6	5)‾9 2 5	7)‾7 9 1
4.	9)‾9 6 9	6)‾7 9 7	8)‾9 5 3	7)‾8 9 9	5)‾8 6 9

Lesson 6   Problem Solving

Solve each problem.

1. A rocket used 945 kilograms of fuel in nine seconds during liftoff. The same amount of fuel was used each second. How many kilograms of fuel were used each second?

_____ kilograms were used each second.

2. A team for a relay race consists of four members who run the same distance. How far would each member run in an 880-yard relay?

Each team member would run _____ yards.

3. There are 435 folding chairs in all. How many rows of 9 chairs each can be formed? How many chairs will be left over?

_____ rows can be formed.

_____ chairs will be left over.

4. Eight boxes of freight weigh 920 kilograms. Each box holds the same amount of freight. How much does each box weigh?

Each box weighs _____ kilograms.

5. A person's weight on Earth is six times more than on the moon. How much would a person who weighs 180 pounds on Earth weigh on the moon?

The person would weigh _____ pounds.

6. There are 627 items to be packed. The items are to be packed 6 to a box. How many boxes can be filled? How many items will be left over?

_____ boxes will be filled.

_____ items will be left over.

7. The weight of four players on a team is 292 kilograms. Suppose each player weighs the same amount. How much would each player weigh?

Each player would weigh _____ kilograms.

1.	
2.	**3.**
4.	**5.**
6.	**7.**

CHAPTER 9 PRACTICE TEST
Division (2- and 3-digit by 1-digit)

Divide.

	a	b	c	d	e
1.	4)2̄6̄	6)3̄9̄	7)6̄8̄	8)8̄8̄	2)8̄6̄
2.	5)9̄0̄	6)7̄8̄	3)6̄8̄	4)8̄7̄	8)9̄8̄
3.	2)1̄0̄4̄	6)2̄4̄6̄	3)2̄1̄9̄	5)2̄8̄5̄	8)6̄7̄2̄
4.	4)3̄6̄3̄	3)2̄7̄8̄	5)4̄2̄7̄	9)6̄2̄7̄	8)4̄6̄5̄
5.	9)9̄8̄1̄	4)8̄9̄2̄	7)9̄5̄2̄	4)8̄4̄3̄	8)9̄8̄6̄

CHAPTER 10 PRETEST
Division (4-digit by 1-digit)

Divide.

	a	*b*	*c*	*d*
1.	3)2 4 9	5)4 5 2	8)4 8 9 3	6)4 2 5 7
2.	4)8 4 0 8	9)9 0 8 1	7)9 1 4 7	5)6 7 2 4
3.	2)1 2 6 8	4)3 2 8 2	6)2 4 1 6	9)1 7 6 4
4.	5)5 6 5 0	3)3 6 9 2	7)8 4 3 5	8)9 4 7 2

Lesson 1 Division (3-digit)

Think about each digit in the quotient.

hundreds

$3\overline{)4}$

```
    1
3) 427
  300
  127
```

tens

$3\overline{)12}$

```
   14
3) 427
  300
  127
  120
    7
```

ones

$3\overline{)7}$

```
  142 r1
3) 427
  300
  127
  120
    7
    6
    1
```

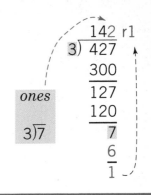

Divide.

	a	b	c	d	e

1. $2\overline{)268}$ $3\overline{)459}$ $5\overline{)705}$ $6\overline{)965}$ $2\overline{)483}$

2. $6\overline{)555}$ $9\overline{)421}$ $7\overline{)686}$ $8\overline{)406}$ $3\overline{)254}$

3. $7\overline{)476}$ $3\overline{)654}$ $5\overline{)982}$ $9\overline{)734}$ $4\overline{)502}$

Lesson 2 Division (4-digit)

Think about each digit in the quotient.

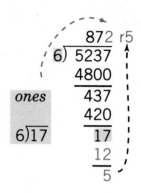

hundreds

$6)\overline{52}$

$\begin{array}{r} 8 \\ 6)\overline{5237} \\ 4800 \\ \hline 437 \end{array}$

tens $\begin{array}{r} 87 \\ 6)\overline{5237} \\ 4800 \\ \hline 437 \\ 420 \\ \hline 17 \end{array}$

$6)\overline{43}$

ones $6)\overline{17}$

$\begin{array}{r} 872 \ r5 \\ 6)\overline{5237} \\ 4800 \\ \hline 437 \\ 420 \\ \hline 17 \\ 12 \\ \hline 5 \end{array}$

Divide.

	a	b	c	d	e

1. $2)\overline{1248}$ $7)\overline{4291}$ $3)\overline{2580}$ $6)\overline{3348}$ $5)\overline{3745}$

2. $8)\overline{6483}$ $9)\overline{8174}$ $4)\overline{2509}$ $2)\overline{1983}$ $7)\overline{3490}$

3. $3)\overline{2922}$ $6)\overline{5277}$ $5)\overline{4350}$ $8)\overline{6543}$ $9)\overline{6255}$

Lesson 3 Division (4-digit)

Think about each digit in the quotient.

thousands

$4\overline{)9}$

$\begin{array}{r} 2 \\ 4\overline{)\ 9527} \\ 8000 \\ \hline 1527 \end{array}$

hundreds $4\overline{)15}$

$\begin{array}{r} 23 \\ 4\overline{)\ 9527} \\ 8000 \\ \hline 1527 \\ 1200 \\ \hline 327 \end{array}$

tens $4\overline{)32}$

$\begin{array}{r} 238 \\ 4\overline{)\ 9527} \\ 8000 \\ \hline 1527 \\ 1200 \\ \hline 327 \\ 320 \\ \hline 7 \end{array}$

ones $4\overline{)7}$

$\begin{array}{r} 2381\ \text{r3} \\ 4\overline{)\ 9527} \\ 8000 \\ \hline 1527 \\ 1200 \\ \hline 327 \\ 320 \\ \hline 7 \\ 4 \\ \hline 3 \end{array}$

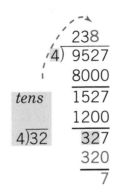
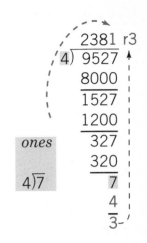

Divide.

	a	b	c	d	e
1.	$6\overline{)6684}$	$3\overline{)9642}$	$8\overline{)8032}$	$5\overline{)7845}$	$9\overline{)9819}$
2.	$5\overline{)5758}$	$4\overline{)4287}$	$7\overline{)9466}$	$6\overline{)8755}$	$4\overline{)9979}$

CHAPTER 10

Lesson 3   Problem Solving

Solve each problem.

1. A trip is 956 kilometers long. The same distance will be driven on each of two days. How many kilometers will be driven on each day?

   _____ kilometers will be driven each day.

2. A pony-express trip of 1,230 miles took five days. The rider went the same number of miles each day. How many miles did the rider go each day?

   The rider went _____ miles each day.

3. The Thompsons' car cost $9,624. They paid the same amount in each of three years. How much did they pay each year?

   _____ was paid each year.

4. A tank contains 555 liters of oil. Nine liters of oil are used each day. How many days will the supply last? How many liters will be left over?

   The supply will last _____ days.

   _____ liters will be left over.

5. At Pointview School there are four grades and 4,196 students. There is the same number of students in each grade. How many students are in each grade?

   _____ students are in each grade.

6. There are 5,280 feet in a mile and 3 feet in a yard. How many yards are there in a mile?

   _____ yards are in a mile.

7. Each box holds 8 cans. There are 9,539 cans to be packed. How many boxes will be filled? How many cans will be left over?

   _____ boxes will be filled.

   _____ cans will be left over.

1.	
2.	**3.**
4.	**5.**
6.	**7.**

CHAPTER 10 PRACTICE TEST
Division (4-digit)

Divide.

	a	*b*	*c*	*d*	*e*
1.	$3\overline{)216}$	$6\overline{)481}$	$4\overline{)2284}$	$2\overline{)1256}$	$6\overline{)1848}$
2.	$7\overline{)5478}$	$8\overline{)3645}$	$5\overline{)4378}$	$8\overline{)6245}$	$4\overline{)3675}$
3.	$9\overline{)9819}$	$7\overline{)8442}$	$5\overline{)7605}$	$5\overline{)5256}$	$7\overline{)9184}$
4.	$3\overline{)8252}$	$6\overline{)8755}$	$8\overline{)9147}$	$9\overline{)9171}$	$3\overline{)8678}$

CHAPTER 10

CHAPTER 11 PRETEST
Multiplication and Division

Multiply or divide. Check each answer.

	a	*b*	*c*
1.	2 1 3 ×2	2 2 0 2 ×4	3 1 2 3 ×3
2.	7 8 7 ×8	1 6 5 4 ×5	3 5 9 ×6
3.	2)1 2 6	3)6 7 9	8)9 6 4
4.	5)4 5 0 7	9)8 7 2 2	6)6 7 2 6
5.	4)1 3 4 5	7)1 2 3 4	5)6 7 5 2

Lesson 1   Multiplication and Division

$$
\begin{array}{c}
19 \longrightarrow \quad 19 \\
\times 5 \longrightarrow 5)\overline{95} \\
\hline 95 \dashrightarrow
\end{array}
\qquad
\begin{array}{c}
507 \\
\times 8 \\
\hline 4056
\end{array}
\text{ so } 8)\overline{4056}
\qquad
\begin{array}{c}
13 \longrightarrow 13 \\
7)\overline{91} \quad \times 7 \\
\underline{70} \quad \overline{91} \\
\overline{21} \\
\underline{21} \\
\overline{0}
\end{array}
\qquad
\begin{array}{c}
272 \\
7)\overline{1904}
\end{array}
\text{ so }
\begin{array}{c}
272 \\
\times 7
\end{array}
$$

Complete the following.

	a		b

1.
$$
\begin{array}{c}
117 \\
\times 5 \\
\hline 585
\end{array}
\text{ so } 5)\overline{585}
\qquad\qquad
3)\overline{285}
\begin{array}{c}
95 \\
\end{array}
\text{ so }
\begin{array}{c}
95 \\
\times 3
\end{array}
$$

2.
$$
\begin{array}{c}
219 \\
\times 7 \\
\hline 1533
\end{array}
\text{ so } 7)\overline{1533}
\qquad\qquad
9)\overline{1827}
\begin{array}{c}
203 \\
\end{array}
\text{ so }
\begin{array}{c}
203 \\
\times 9
\end{array}
$$

Multiply or divide.

	a		b		c		d

3.
$$
\begin{array}{c}
27 \\
\times 5
\end{array}
\qquad
5)\overline{135}
\qquad
\begin{array}{c}
123 \\
\times 8
\end{array}
\qquad
8)\overline{984}
$$

4.
$$
\begin{array}{c}
728 \\
\times 6
\end{array}
\qquad
6)\overline{4368}
\qquad
4)\overline{8324}
\qquad
\begin{array}{c}
2081 \\
\times 4
\end{array}
$$

5.
$$
9)\overline{2214}
\qquad
\begin{array}{c}
246 \\
\times 9
\end{array}
\qquad
3)\overline{561}
\qquad
\begin{array}{c}
187 \\
\times 3
\end{array}
$$

Lesson 1   Problem Solving

Solve each problem.

1. Charles worked 80 problems in five days. He worked the same number of problems each day. How many problems did he work each day?

   He worked _____ problems each day.

2. There are nine rows of tiles on a floor. There are 18 tiles in each row. How many tiles are there on the floor?

   _____ tiles are on the floor.

3. An eight-story apartment building is 112 feet high. Each story is the same height. What is the height of each story?

   Each story is _____ feet high.

4. A sheet of plywood weighs 21 kilograms. What would be the weight of eight sheets?

   The weight would be _____ kilograms.

5. A company made 4,325 cars in five days. The same number of cars was made each day. How many cars were made each day?

   _____ cars were made each day.

6. The seating capacity of a sports arena is 7,560. The seats are arranged in six sections of the same size. How many seats are there in each section?

   _____ seats are in each section.

7. The school library has 8,096 books. The same number of books is stored along each of the four walls. How many books are along each wall?

   _____ books are along each wall.

1.	
2.	**3.**
4.	**5.**
6.	**7.**

Lesson 2   Checking Multiplication

These should
be the same. --------→ 321

321 ↙   *Check*   3) 963
×3   900
‾‾‾   ‾‾‾
963   63
                            60
                            ‾‾
To check 3 × 321 = 963,   3
divide 963 by 3.   3
                            ‾‾
                             0

These should
be the same. --------→ 1243

1243 ↙   *Check*   4) 4972
×4   4000
‾‾‾‾   ‾‾‾‾
4972   972
                            800
                            ‾‾‾
To check 4 × 1243 = 4,972, 172
                            160
divide _____ by ___.   ‾‾‾
                             12
                             12
                            ‾‾
                             0

Multiply. Check each answer.

	a	*b*	*c*
1.	2 3 1 ×3	6 7 8 ×7	9 7 5 ×8
2.	1 2 3 4 ×2	2 6 7 5 ×9	1 2 5 7 ×6
3.	2 3 0 2 ×4	7 5 8 2 ×5	3 5 4 3 ×7

CHAPTER 11

Lesson 2   Problem Solving

Solve each problem. Check each answer.

1. Each of the four members of a relay team runs 440 yards. What is the total distance the team will run?

   The team will run _____ yards.

2. Benjamin delivers 165 papers each day. How many papers does he deliver in a week?

   He delivers _____ papers in a week.

3. There are 125 nails in a 1-pound pack. How many nails will be in a 5-pound pack?

   _____ nails will be in a 5-pound pack.

4. It takes 1,200 trading stamps to fill a book. How many stamps will it take to fill six books?

   It will take _____ stamps.

5. A contractor estimated that it would take 2,072 bricks to build each of the four walls of a new house. About how many bricks would it take to build all four walls?

   It would take about _____ bricks.

6. There are seven boxes on a truck. Each box weighs 650 kilograms. What is the weight of all the boxes?

   The weight of all the boxes is _____ kilograms.

7. Each box in problem 6 has a value of $845. What is the value of all the boxes?

   The value is $_____.

8. Miss Brooks travels 1,025 kilometers each week. How many kilometers will she travel in six weeks?

   She will travel _____ kilometers.

1.	2.
3.	4.
5.	6.
7.	8.

Lesson 3 Checking Division

```
      442
   3) 1326          Check    442
      1200                    ×3
      ────                  ─────
       126   These should ----→ 1326
       120   be the same.
       ───
         6
         6
       ───
         0
              To check 1326 ÷ 3 = 442,
              multiply 442 by 3.
```

```
       3453
   5) 17265          Check   3453
      15000                   ×5
      ─────                 ──────
       2265   These should ----→ 17265
       2000   be the same.
       ────
        265
        250
        ───
         15   To check 17265 ÷ 5 = 3453,
         15
        ───
          0   multiply 3,453 by ____.
```

Divide. Check each answer.

	a	b	c

1. 3) 2 4 6   5) 6 7 5   9) 9 8 1

2. 9) 1 5 6 6   7) 7 8 4 7   4) 6 2 5 6

3. 3) 1 7 1 0   7) 1 4 3 5   2) 9 8 5 8

CHAPTER 11

Lesson 3 Problem Solving

Solve each problem. Check each answer.

1. A plane traveled 900 kilometers in two hours. The same distance was traveled each hour. How far did the plane travel each hour?

   _____ kilometers were traveled each hour.

2. A lunchroom served 960 lunches in three hours. The same number of lunches was served each hour. How many lunches were served each hour?

   _____ lunches were served each hour.

3. A company has 7,200 workers. There are eight plants with the same number of workers at each plant. How many workers are at each plant?

   _____ workers are at each plant.

4. A school ordered 2,688 books. The books will be delivered in seven equal shipments. How many books will be in each shipment?

   _____ books will be in each shipment.

5. A shipping company had 5,900 kilograms of freight delivered in five loads. Each load had the same weight. Find the weight of each load.

   The weight of each load was _____ kilograms.

6. There are 5,040 calories in six cups of roasted peanuts. How many calories are there in one cup of roasted peanuts?

   _____ calories are in one cup of peanuts.

7. A refinery filled 8,652 oil cans in three hours. The same number of cans was filled each hour. How many cans were filled each hour?

   _____ cans were filled each hour.

1.	
2.	**3.**
4.	**5.**
6.	**7.**

Lesson 4   Checking Division

```
        263 r4
   7) 1845
       14
      ---
      445
       42
      ---
       25
       21
       --
        4
```

Check   263
              ×7

            1841
              +4

These should 1845
be the same. →

To check the division,
multiply 263 by 7 and
then add the remainder 4.

Divide. Check each answer.

| *a* | *b* | *c* |

1. 2) 1 5 7   5) 6 7 9   4) 8 0 7

2. 3) 1 3 4 9   6) 7 8 5 5   7) 9 4 6 3

3. 9) 1 2 3 4   8) 9 6 5 4   6) 9 7 8 5

Lesson 4 Problem Solving

Solve each problem. Check each answer.

1. A marching band has 126 members. How many rows of 8 members each can be formed? How many members will be in the next row?

_____ rows can be formed.

_____ members will be in the next row.

1.

2. Mr. Carpenter has 228 floor tiles. How many rows of 9 tiles each can he lay? How many tiles will be left over?

He can lay _____ rows of 9 tiles each.

_____ tiles will be left over.

2.

3. Seven pirates want to share 1,006 coins so that each will get the same number of coins. How many coins will each pirate get? How many coins will be left over?

Each pirate will get _____ coins.

_____ coins will be left over.

3.

4. There will be 1,012 people at a large banquet. Six people will sit at each table. How many tables will be filled? How many people will sit at a table that is not completely filled?

_____ tables will be filled.

_____ people will sit at a table that is not completely filled.

4.

5. Last week 2,407 empty bottles were returned to a store. How many cartons of 8 bottles each could be filled? How many bottles would be left over?

_____ cartons could be filled.

_____ bottles would be left over.

5.

Lesson 5 Estimation (products)

Estimate 27 × 5.

Round 27 to its highest place value. Then, multiply by 5.

$$27 \longrightarrow 30$$
$$\times\ 5 \qquad \times\ 5$$
$$150\ (3 \times 5 = 15 \text{ and then write the zero})$$

27 × 5 is about 150.

Estimate each answer.

	a	b	c	d	e
1.	57 × 6	23 × 9	45 × 8	62 × 3	97 × 9
2.	21 × 2	72 × 7	57 × 4	93 × 6	38 × 5
3.	183 × 2	786 × 8	359 × 7	223 × 6	627 × 9
4.	685 × 5	428 × 9	262 × 4	219 × 3	812 × 7
5.	2082 × 3	7563 × 9	8235 × 6	6752 × 5	4913 × 8
6.	7921 × 9	8339 × 3	4275 × 8	2712 × 9	3657 × 6

CHAPTER 11

Orange Book

Lesson 5 Problem Solving

Solve each problem.

1. A grocer ordered eight cartons of cookies. There are 18 boxes in each carton. About how many boxes of cookies did he order?

   about _____ boxes

2. Over the summer, Joe went on a long bike ride 34 times. Each time he rode 8 miles. About how far did he ride his bike in all?

   about _____ miles

3. Six classes of students signed up for the Fun Run. Each class has 22 students. About how many students signed up to run?

   about _____ students

4. For a fund-raiser, Prairie View School District collected box tops. The goal was for each of the 628 students to bring in eight box tops. If everyone brought in eight box tops, about how many would the district have?

   about _____ box tops

5. Eight girls stuffed envelopes for a campaign. They each filled 415 envelopes. About how many envelopes did they fill in all?

   about _____ envelopes

6. Each grandchild in Mrs. Wheeler's family saved $7 so that they could buy Grandma Jean a special gift. There are 13 grandchildren in the family. About how much money did they save?

   about _____

7. Doris's Dishwasher Depot had a big sale. She sold eight dishwashers in one day. Each one sold for $685. About how much money did she collect at her sale?

   about _____

	1.
	2.
	3.
	4.
	5.
	6.
	7.

Lesson 6   Estimation (quotients)

Estimate $9\overline{)8078}$.

Since $9\overline{)81}$ is a familiar division fact, round 8,078 to the nearest hundred, 8,100.

$$\begin{array}{c} 900 \\ 9\overline{)8100} \end{array} \quad (81 \div 9 = 9 \text{ and then write the zeros})$$

$9\overline{)8078} \longrightarrow 9\overline{)8100}$

$9\overline{)8078}$ is about 900.

Estimate each answer.

	a	b	c	d	e
1.	$2\overline{)104}$	$9\overline{)266}$	$7\overline{)629}$	$9\overline{)362}$	$6\overline{)239}$
2.	$4\overline{)123}$	$8\overline{)639}$	$6\overline{)419}$	$2\overline{)125}$	$3\overline{)256}$
3.	$2\overline{)147}$	$3\overline{)215}$	$6\overline{)335}$	$3\overline{)191}$	$7\overline{)227}$
4.	$3\overline{)1819}$	$9\overline{)2695}$	$3\overline{)1819}$	$9\overline{)2695}$	$3\overline{)8755}$
5.	$5\overline{)5254}$	$7\overline{)4896}$	$6\overline{)2321}$	$5\overline{)4459}$	$6\overline{)3728}$
6.	$9\overline{)7122}$	$8\overline{)2691}$	$8\overline{)1562}$	$8\overline{)4721}$	$5\overline{)2682}$

Lesson 6 Problem Solving

Solve each problem.

1. Miss G, the science teacher, gave her seven students 136 tadpoles to observe. About how many tadpoles did each student receive?

   about _____ tadpoles

2. At the swimming party, 318 students were divided among the eight pools. About how many students swam in each pool?

   about _____ students

3. After Daniel baked nine batches of cookies, he had 454 cookies. About how many cookies were made in each batch?

   about _____ cookies

4. Lien divided 729 pennies into nine equal piles. About how many pennies were in each pile?

   about _____ pennies

5. Nine teachers had 1,910 tests to grade. About how many tests will each teacher grade?

   about _____ tests

6. The school library has 1,585 books that are placed on shelves. If there are eight shelves in the library, about how many books are on one shelf?

   about _____ books

7. At a concert, 2,698 screaming fans wanted the autographs of the nine singers. If each fan gets only one autograph, about how many autographs did each singer sign?

   about _____ autographs

1.
2.
3.
4.
5.
6.
7.

CHAPTER 11 PRACTICE TEST
Multiplication and Division

Multiply or divide. Check each answer.

	a	*b*	*c*
1.	2 1 3 ×3	6 2 1 0 ×4	8 3 2 0 ×4
2.	3 2 4 5 ×6	1 3 4 2 ×7	6 8 7 1 ×5
3.	6)2 7 6	8)1 6 4 0	6)3 4 4
4.	6)5 6 5	3)1 4 5 7	5)8 3 7 1

Estimate each answer.

	a	*b*	*c*
5.	7 7 4 ×3	8 1 2 5 ×4	5 9 7 8 ×8
6.	7)5 5 7	7)1 3 8 1	5)3 4 6 7

7. Each page of a math book contained eight problems. There were 184 pages in the book.

Use estimation to find about how many problems there are in the book.

about _____ problems

Now find the exact answer. How many problems are there in the book?

_____ total problems

7.

CHAPTER 12 PRETEST
Fractions

What fraction of each figure is shaded?

1.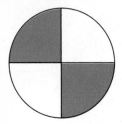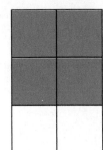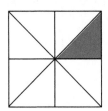

_____   _____   _____   _____

What fraction of each set is shaded?

2.   

_____   _____   _____   _____

Find the number that makes the two fractions equivalent.

3. $\dfrac{1}{8} = \dfrac{}{16}$ $\dfrac{2}{3} = \dfrac{12}{}$ $\dfrac{8}{12} = \dfrac{2}{}$ $\dfrac{3}{6} = \dfrac{}{18}$

Add.

	a	b	c	d

4. $\dfrac{9}{15} + \dfrac{5}{15} =$ $\dfrac{4}{9} + \dfrac{2}{9} =$ $\dfrac{1}{6} + \dfrac{2}{6} =$ $\dfrac{4}{10} + \dfrac{3}{10} =$

Subtract.

5. $\dfrac{4}{5} - \dfrac{2}{5} =$ $\dfrac{6}{7} - \dfrac{3}{7} =$ $\dfrac{2}{6} - \dfrac{1}{6} =$ $\dfrac{8}{9} - \dfrac{3}{9} =$

Lesson 1 Fractions of a Whole

What fraction of the figure is shaded?

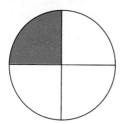

The circle is cut into four equal pieces, so the denominator, or the bottom of the fraction, is 4.

Only one part is shaded, so the numerator, or the top of the fraction, is 1.

$\frac{1}{4}$ of the figure is shaded.

What fraction of the figure is shaded?

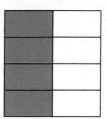

$\frac{4}{8}$ ← parts shaded—numerator.
  ← parts in all—denominator.

$\frac{4}{8}$ of the figure is shaded.

What fraction of each figure is shaded?

| | *a* | *b* | *c* |

1.

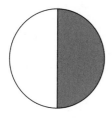 $\frac{}{2}$

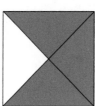 $\frac{3}{}$

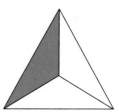 $\frac{}{3}$

2.

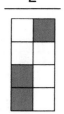

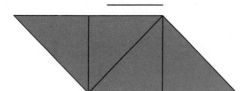

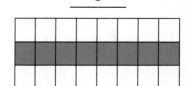

3.

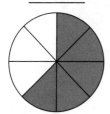

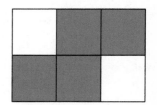

Lesson 2 Fractions of a Set

What fraction of the set is shaded?

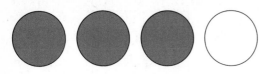

$\dfrac{3}{4}$ ← objects shaded—numerator.
← object in all—denominator.

$\dfrac{3}{4}$ of the set is shaded.

What fraction of the set is shaded?

$\dfrac{2}{3}$ ← number shaded
← number in all

$\dfrac{2}{3}$ of the set is shaded.

What fraction of each set is shaded?

a b

1.

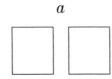

$\dfrac{1}{\rule{1cm}{0.4pt}}$ $\dfrac{2}{\rule{1cm}{0.4pt}}$

2.

$\dfrac{\rule{1cm}{0.4pt}}{\rule{1cm}{0.4pt}}$ $\dfrac{\rule{1cm}{0.4pt}}{\rule{1cm}{0.4pt}}$

3.

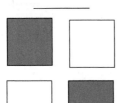

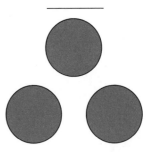

$\dfrac{\rule{1cm}{0.4pt}}{\rule{1cm}{0.4pt}}$ $\dfrac{\rule{1cm}{0.4pt}}{\rule{1cm}{0.4pt}}$

Lesson 3 Equivalent Fractions (multiply)

Find an equivalent fraction.

$\frac{2}{3} = \frac{?}{?}$

You can find an equivalent fraction for $\frac{2}{3}$ if you multiply both the numerator and the denominator by the same number, such as 2.

$\frac{2}{3} = \frac{2 \times 2}{3 \times 2} = \frac{4}{6}$ Multiply the numerator by 2 to get 4.
Multiply the denominator by 2 to get 6.

$\frac{2}{3} = \frac{4}{6}$ $\frac{2}{3}$ and $\frac{4}{6}$ are equivalent fractions.

Use multiplication to find each equivalent fraction.

	a	b	c
1.	$\frac{1}{5} = \frac{1 \times 2}{5 \times 2} = \frac{2}{\quad}$	$\frac{2}{4} = \frac{2 \times 3}{4 \times 3} = \frac{}{12}$	$\frac{1}{3} = \frac{1 \times 4}{3 \times 4} = \frac{4}{\quad}$
2.	$\frac{2}{3} = \frac{2 \times 3}{3 \times} = \frac{}{}$	$\frac{5}{6} = \frac{5 \times}{6 \times 2} = \frac{}{}$	$\frac{1}{5} = \frac{1 \times 2}{\times 2} = \frac{}{}$
3.	$\frac{2}{3} = \frac{2 \times 5}{3 \times} = \frac{}{}$	$\frac{4}{6} = \frac{4 \times}{6 \times 2} = \frac{}{}$	$\frac{1}{3} = \frac{1 \times 3}{3 \times} = \frac{}{}$
4.	$\frac{1}{4} = \frac{1 \times}{4 \times} = \frac{}{16}$	$\frac{1}{3} = \frac{1 \times}{3 \times} = \frac{5}{}$	$\frac{3}{4} = \frac{3 \times}{4 \times} = \frac{}{8}$

	a	b	c	d
5.	$\frac{3}{5} = \frac{}{15}$	$\frac{6}{8} = \frac{}{16}$	$\frac{2}{3} = \frac{}{27}$	$\frac{1}{4} = \frac{}{8}$
6.	$\frac{4}{5} = \frac{}{20}$	$\frac{5}{7} = \frac{10}{}$	$\frac{1}{6} = \frac{}{12}$	$\frac{8}{9} = \frac{}{27}$

Lesson 4 Equivalent Fractions (divide)

Find an equivalent fraction.

$\dfrac{8}{10} = \dfrac{?}{?}$

You can find an equivalent fraction for $\dfrac{8}{10}$ if you can divide both the numerator and the denominator by the same number, such as 2.

$\dfrac{8}{10} = \dfrac{8 \div 2}{10 \div 2} = \dfrac{4}{5}$

Divide the numerator by 2 to get 4.
Divide the denominator by 2 to get 5.

$\dfrac{8}{10} = \dfrac{4}{5}$

$\dfrac{8}{10}$ and $\dfrac{4}{5}$ are equivalent fractions.

Use division to find each equivalent fraction.

a	b	c
1. $\dfrac{10}{12} = \dfrac{10 \div 2}{12 \div 2} = \dfrac{5}{\quad}$	$\dfrac{4}{8} = \dfrac{4 \div 2}{8 \div 2} = \dfrac{\quad}{4}$	$\dfrac{4}{12} = \dfrac{4 \div 4}{12 \div 4} = \dfrac{1}{\quad}$
2. $\dfrac{5}{10} = \dfrac{5 \div 5}{10 \div} = \dfrac{\quad}{\quad}$	$\dfrac{6}{12} = \dfrac{6 \div}{12 \div 6} = \dfrac{\quad}{\quad}$	$\dfrac{6}{9} = \dfrac{6 \div 3}{\quad \div 3} = \dfrac{\quad}{\quad}$
3. $\dfrac{12}{15} = \dfrac{12 \div 3}{15 \div} = \dfrac{\quad}{\quad}$	$\dfrac{4}{10} = \dfrac{4 \div}{10 \div 2} = \dfrac{\quad}{\quad}$	$\dfrac{12}{16} = \dfrac{12 \div 4}{16 \div} = \dfrac{\quad}{\quad}$
4. $\dfrac{6}{15} = \dfrac{6 \div}{15 \div} = \dfrac{\quad}{5}$	$\dfrac{10}{15} = \dfrac{10 \div}{15 \div} = \dfrac{2}{\quad}$	$\dfrac{4}{16} = \dfrac{4 \div}{16 \div} = \dfrac{\quad}{4}$

a	b	c	d
5. $\dfrac{12}{16} = \dfrac{\quad}{8}$	$\dfrac{2}{6} = \dfrac{\quad}{3}$	$\dfrac{6}{18} = \dfrac{\quad}{6}$	$\dfrac{10}{16} = \dfrac{\quad}{8}$
6. $\dfrac{8}{20} = \dfrac{\quad}{10}$	$\dfrac{6}{8} = \dfrac{3}{\quad}$	$\dfrac{2}{12} = \dfrac{\quad}{6}$	$\dfrac{8}{12} = \dfrac{\quad}{3}$

Lesson 5 Adding Fractions (like denominators)

$\dfrac{1}{4} + \dfrac{2}{4} = \dfrac{?}{?}$

$\dfrac{1}{4} + \dfrac{2}{4} = \dfrac{1+2}{4}$ Add the numerators.

$\qquad = \dfrac{3}{4}$ Write the sum over the common denominator.

Add the fractions.

	a	*b*	*c*
1.	$\dfrac{1}{3} + \dfrac{1}{3} = \dfrac{}{3}$	$\dfrac{2}{5} + \dfrac{1}{5} = \dfrac{}{5}$	$\dfrac{2}{4} + \dfrac{1}{4} = \dfrac{3}{}$
2.	$\dfrac{3}{7} + \dfrac{2}{7} = -$	$\dfrac{2}{9} + \dfrac{3}{9} = -$	$\dfrac{1}{7} + \dfrac{2}{7} = -$
3.	$\dfrac{3}{10} + \dfrac{3}{10} = -$	$\dfrac{3}{8} + \dfrac{1}{8} = -$	$\dfrac{2}{5} + \dfrac{1}{5} = -$

	a	*b*	*c*	*d*	*e*
4.	$\dfrac{1}{6}$ $+\dfrac{4}{6}$ $\overline{\dfrac{}{6}}$	$\dfrac{2}{6}$ $+\dfrac{3}{6}$ $\overline{\dfrac{}{6}}$	$\dfrac{1}{8}$ $+\dfrac{3}{8}$ $\overline{\dfrac{}{8}}$	$\dfrac{1}{12}$ $+\dfrac{6}{12}$ $\overline{\dfrac{}{12}}$	$\dfrac{4}{7}$ $+\dfrac{3}{7}$ $\overline{\dfrac{}{7}}$
5.	$\dfrac{3}{5}$ $+\dfrac{1}{5}$	$\dfrac{2}{9}$ $+\dfrac{2}{9}$	$\dfrac{1}{3}$ $+\dfrac{1}{3}$	$\dfrac{4}{10}$ $+\dfrac{5}{10}$	$\dfrac{2}{8}$ $+\dfrac{4}{8}$

Lesson 5 Problem Solving

Solve each problem.

1. It rained $\frac{1}{10}$ of an inch yesterday and $\frac{6}{10}$ of an inch today. How much has it rained in the last two days?

   It has rained _____ of an inch in the last two days.

2. Yu Luan ran $\frac{2}{8}$ of a mile yesterday and $\frac{3}{8}$ of a mile today. What was the total distance she ran in two days?

   Yu Luan ran _____ of a mile in two days.

3. It took Jenna $\frac{1}{6}$ of an hour to water her flowers and $\frac{4}{6}$ of an hour to weed her garden. How long did she work in her yard?

   Jenna worked in her yard _____ of an hour.

4. Tamara is making a submarine sandwich. She bought $\frac{1}{4}$ of a pound of salami and $\frac{2}{4}$ of a pound of ham. How much meat did she buy for her sandwich?

   Tamara bought _____ of a pound of meat.

5. Mrs. Wagner had $\frac{3}{5}$ of a yard of material at home. She went shopping and bought another $\frac{1}{5}$ yard. Now how much material does she have?

   Mrs. Wagner has _____ of a yard of material.

6. Robert read $\frac{3}{7}$ of his book yesterday and $\frac{2}{7}$ of his book today. How much of the book has he read?

   Robert has read _____ of his book.

1.

2.

3.

4.

5.

6.

Lesson 6 Subtracting Fractions (like denominators)

Subtract. $\dfrac{4}{5} - \dfrac{3}{5} = \dfrac{?}{?}$

$\dfrac{4}{5} - \dfrac{3}{5} = \dfrac{4-3}{5}$ Subtract the numerators. Write the difference over the common denominator.

$\dfrac{4-3}{5} = \dfrac{1}{5}$ The difference between $\frac{4}{5}$ and $\frac{3}{5}$ is $\frac{1}{5}$.

$\begin{array}{r} \dfrac{4}{5} \\ -\dfrac{3}{5} \\ \hline \dfrac{1}{5} \end{array}$

Subtract the fractions.

	a	b	c
1.	$\dfrac{6}{10} - \dfrac{2}{10} = \dfrac{}{10}$	$\dfrac{2}{5} - \dfrac{1}{5} = \dfrac{}{5}$	$\dfrac{2}{4} - \dfrac{1}{4} = \dfrac{1}{}$
2.	$\dfrac{3}{5} - \dfrac{2}{5} = -$	$\dfrac{3}{8} - \dfrac{1}{8} = -$	$\dfrac{8}{8} - \dfrac{5}{8} = -$
3.	$\dfrac{4}{5} - \dfrac{3}{5} = -$	$\dfrac{7}{9} - \dfrac{6}{9} = -$	$\dfrac{4}{5} - \dfrac{1}{5} = -$

	a	b	c	d	e
4.	$\begin{array}{r}\dfrac{9}{12}\\-\dfrac{4}{12}\\\hline \dfrac{}{12}\end{array}$	$\begin{array}{r}\dfrac{6}{6}\\-\dfrac{3}{6}\\\hline \dfrac{}{6}\end{array}$	$\begin{array}{r}\dfrac{4}{7}\\-\dfrac{1}{7}\\\hline \dfrac{}{7}\end{array}$	$\begin{array}{r}\dfrac{7}{10}\\-\dfrac{2}{10}\\\hline \dfrac{}{10}\end{array}$	$\begin{array}{r}\dfrac{4}{7}\\-\dfrac{3}{7}\\\hline \dfrac{}{7}\end{array}$
5.	$\begin{array}{r}\dfrac{6}{8}\\-\dfrac{2}{8}\\\hline\end{array}$	$\begin{array}{r}\dfrac{9}{11}\\-\dfrac{3}{11}\\\hline\end{array}$	$\begin{array}{r}\dfrac{2}{3}\\-\dfrac{1}{3}\\\hline\end{array}$	$\begin{array}{r}\dfrac{5}{7}\\-\dfrac{2}{7}\\\hline\end{array}$	$\begin{array}{r}\dfrac{8}{10}\\-\dfrac{3}{10}\\\hline\end{array}$

Lesson 6   Problem Solving

Solve each problem.

1. The small beetle was $\frac{1}{5}$ of an inch wide and $\frac{2}{5}$ of an inch long. How much longer was the beetle than it was wide?

   The beetle was _____ of an inch longer than it was wide.

2. Seven-eighths of the theater was full for Friday night's performance. On Saturday night only $\frac{5}{8}$ of the theater was full. How much more of the theater was full on Friday night than on Saturday night?

   The theater was _____ more full on Friday night than on Saturday night.

3. The pool was $\frac{6}{8}$ full last night and $\frac{7}{8}$ full this morning. How much of the pool was filled during the night?

   _____ of the pool was filled last night.

4. Liam always uses $\frac{9}{12}$ of a cup of butter in his cookies. Gloria only uses $\frac{8}{12}$ of a cup. How much more butter does Liam use than Gloria?

   Liam uses _____ of a cup more butter than Gloria.

5. It takes Margaret $\frac{3}{6}$ of an hour to walk to school. It only takes Jordan $\frac{2}{6}$ of an hour to get there. How much longer does it take Margaret to walk to school than it takes Jordan?

   It takes Margaret _____ of an hour longer to walk to school than it takes Jordan.

1.

2.

3.

4.

5.

CHAPTER 12 PRACTICE TEST
Fractions

What fraction of each figure is shaded?

1.

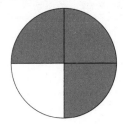 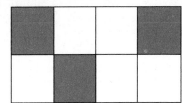 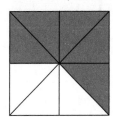 

_____   _____   _____   _____

What fraction of each set is shaded?

2.

  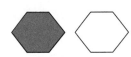

_____   _____   _____   _____

Find the number that makes the two fractions equivalent.

3. $\dfrac{4}{8} = \dfrac{}{16}$ $\dfrac{2}{5} = \dfrac{18}{}$ $\dfrac{1}{4} = \dfrac{3}{}$ $\dfrac{3}{5} = \dfrac{}{15}$

Add.

	a	b	c	d

4. $\dfrac{4}{8} + \dfrac{3}{8} =$ $\dfrac{1}{3} + \dfrac{1}{3} =$ $\dfrac{2}{10} + \dfrac{3}{10} =$ $\dfrac{2}{9} + \dfrac{6}{9} =$

Subtract.

5. $\dfrac{3}{4} - \dfrac{1}{4} =$ $\dfrac{5}{12} - \dfrac{3}{12} =$ $\dfrac{7}{12} - \dfrac{4}{12} =$ $\dfrac{3}{6} - \dfrac{2}{6} =$

CHAPTER 13 PRETEST
Geometry

Identify each angle as *right, acute,* or *obtuse.*

1.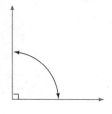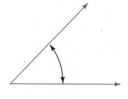

_____   _____   _____

Write the letter for the name of each figure on the blank.

  a   *b*   *c*

2.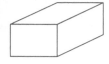

a. sphere
b. quadrilateral
c. octagon
d. rectangular solid
e. cone
f. pentagon

_____   _____   _____

3.

_____   _____   _____

Identify each pair of plane figures as *congruent* or *not congruent.*

  a   *b*   *c*

4.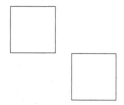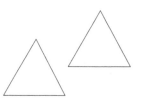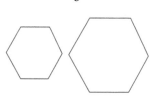

_____   _____   _____

Identify how each figure has been moved by writing *slide, flip,* or *turn.*

5.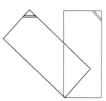

_____   _____   _____

Lesson 1   Angles

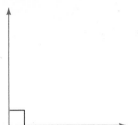

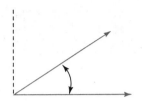

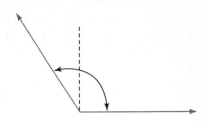

Right angle
90° angle

Acute angle
less than a
right angle

Obtuse angle
more than a right
angle

Identify each angle as *right, acute,* or *obtuse.*

	a	b	c

1.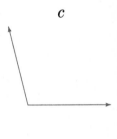

_____   _____   _____

2.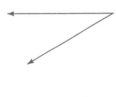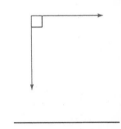

_____   _____   _____

3.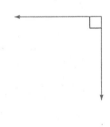

_____   _____   _____

4.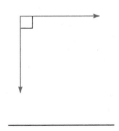

_____   _____   _____

Lesson 1 Problem Solving

Identify the angle of the clock hands as *acute, right,* or *obtuse*.
Then answer each question.

1.

The hands of the clock form a(n) _____ angle.

2.

The hands of the clock form a(n) _____ angle.

3.

The hands of the clock form a(n) _____ angle.

4.

The hands of the clock form a(n) _____ angle.

5.

The hands of the clock form a(n) _____ angle.

Lesson 2 Parallel and Perpendicular

Parallel lines never meet. They are always the same distance apart.

Perpendicular lines intersect, or cross each other, to form right angles.

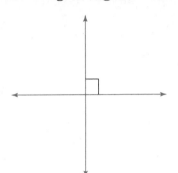

Identify each pair of lines as *parallel, perpendicular,* or *neither.*

	a	*b*	*c*	*d*

1.

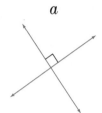 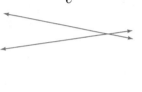 

_____ _____ _____ _____

2.

_____ _____ _____ _____

3.

  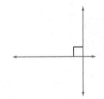

_____ _____ _____ _____

4.

  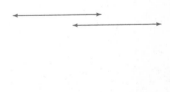

_____ _____ _____ _____

CHAPTER 13

Lesson 2 Problem Solving

Identify the roads as being *parallel*, *perpendicular*, or *neither*.
Then answer each question.

1.

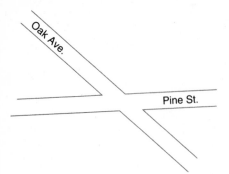

Do the roads cross each other? _____

Do the roads form right angles? _____

The roads form _____ lines.

2.

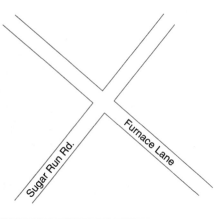

Do the roads form right angles? _____

Do the roads cross each other? _____

The roads form _____ lines.

3.

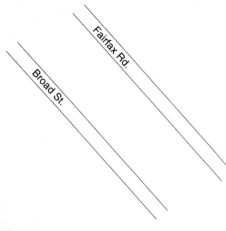

Do the roads cross each other? _____

Do the roads form right angles? _____

The roads form _____ lines.

4.

Do the roads form right angles? _____

Do the roads cross each other? _____

The roads form _____ lines.

Lesson 3 Plane Figures

A plane figure, or polygon, is a closed figure with three or more straight sides. Vertices are the points where two sides meet.

Plane figures are named for the number of sides and vertices.

triangle	**quadrilateral**	**pentagon**
3 sides 3 vertices	4 sides 4 vertices	5 sides 5 vertices
hexagon	**heptagon**	**octagon**
6 sides 6 vertices	7 sides 7 vertices	8 sides 8 vertices

Identify each plane figure as a *triangle, quadrilateral, pentagon, hexagon, heptagon,* or *octagon*.

a   *b*   *c*   *d*

1.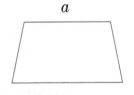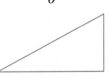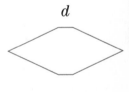

_____ _____ _____ _____

2.

_____ _____ _____ _____

3.

_____ _____ _____ _____

Orange Book

Lesson 4   Solid Figures

Solid figures have three dimensions.

Some solid figures have flat surfaces.

   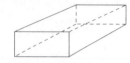   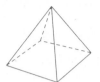

| **cube** | **rectangular solid** | **pyramid** |

Some solid figures have curved surfaces.

| **cone** | **sphere** | **cylinder** |

Identify each object as being closest to the shape of a *cube, rectangular solid, pyramid, cone, sphere,* or *cylinder.*

| | *a* | *b* | *c* | *d* |

1.

_____   _____   _____   _____

2.
   cylinder

_____   _____   _____   _____

3.
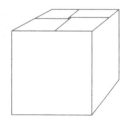      

_____   _____   _____   _____

Lesson 5 Congruence

Two figures that have the same size and shape are **congruent.**

These figures are **congruent.**

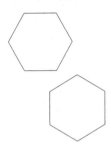

These figures are **not congruent.**

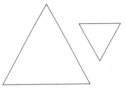

Identify each pair of plane figures as *congruent* or *not congruent.*

| | *a* | *b* | *c* |

1.

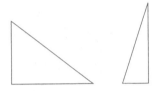   _____

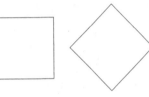   _____

   _____

2.

   _____

(b) _____

(c) _____

3.

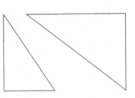   _____

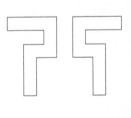   _____

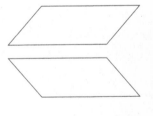   _____

4.

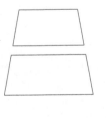   _____

(b) _____

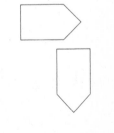   _____

CHAPTER 13

Lesson 6 Symmetry

A plane figure has **symmetry** if it can be folded in one or more ways to make two congruent figures.

This figure has one line of symmetry.	This figure has four lines of symmetry.	This figure has no lines of symmetry.
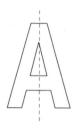	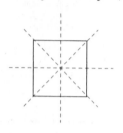	

Draw all the possible lines of symmetry for each figure and write the number of lines. If a figure has no line of symmetry, write *none*.

	a	*b*	*c*
1.			
2.	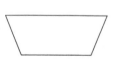 _____	**M** _____	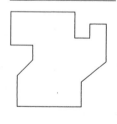 _____
3.	 _____	 _____	_____

Lesson 7 Slides, Flips, and Turns

You can describe how figures have been moved by the terms **slide, flip,** and **turn.**

Slide a figure to move it up, down, left, right, or diagonally.	**Flip** a figure to create a mirror image.	**Turn** a figure to rotate it around a point.

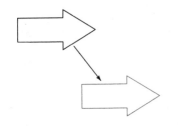 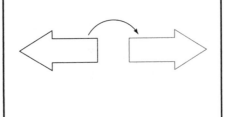 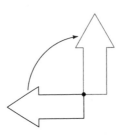

Identify how each figure has been moved by writing *slide, flip,* or *turn.*

        a   *b*   *c*

1.

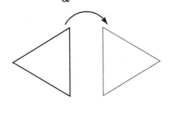   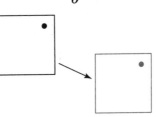   

_____   _____   _____

2.

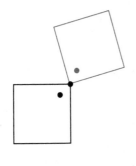   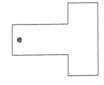

_____   _____   _____

3.

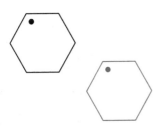   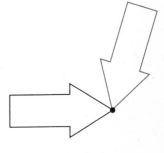

_____   _____   _____

Lesson 7 Problem Solving

What kind of movement is shown by these figures? Answer each question.

1.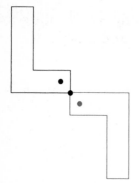

Has the figure moved in one direction? _____

Has the figure been rotated? _____

Has the figure been turned over? _____

The movement is a _____.

2.

Has the figure moved in one direction? _____

Has the figure been rotated? _____

Has the figure been turned over? _____

The movement is a _____.

3.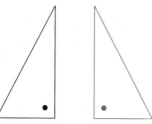

Has the figure moved in one direction? _____

Has the figure been rotated? _____

Has the figure been turned over? _____

The movement is a _____.

What two kinds of movement are shown by these figures?

4.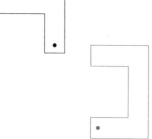

Has the figure moved in one direction? _____

Has the figure been rotated? _____

Has the figure been turned over? _____

The movements are a _____ and a _____.

CHAPTER 13 PRACTICE TEST
Geometry

Identify each angle as *right, acute,* or *obtuse.*

| a | b | c | d |

1.

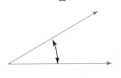  

_____   _____   _____   _____

Identify each plane figure as a *triangle, quadrilateral, pentagon, hexagon, heptagon,* or *octagon.*

2.

  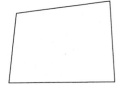 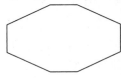

_____   _____   _____   _____

Identify each solid figure as a *cube, rectangular solid, pyramid, cone, sphere,* or *cylinder.*

3.

 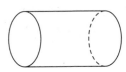 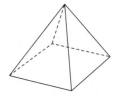

_____   _____   _____   _____

Identify how each figure has been moved by writing *slide, flip,* or *turn.*

4.

 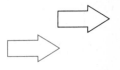

_____   _____   _____

CHAPTER 13

CHAPTER 14 PRETEST
Metric Measurement

Find the length of each line segment to the nearest centimeter (cm).
Then find the length of each line segment to the nearest millimeter (mm).

               a   *b*

1. _____ cm   _____ mm

2. _____ cm   _____ mm

3. _____ cm   _____ mm

Find the perimeter of each figure.

4.

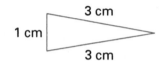

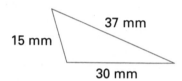

_____ centimeters   _____ millimeters

Complete the following.

               a   *b*

5. 9 centimeters = _____ millimeters   7 kiloliters = _____ liters

6. 4 meters = _____ centimeters   5 grams = _____ milligrams

7. 8 meters = _____ millimeters   13 meters = _____ millimeters

8. 6 liters = _____ milliliters   1 kilometer = _____ meters

9. 9 kilograms = _____ grams   29 liters = _____ milliliters

10. 12 meters = _____ centimeters   23 kiloliters = _____ liters

11. 1 gram = _____ milligrams   80 centimeters = _____ millimeters

12. 92 kilograms = _____ grams   16 kilometers = _____ meters

Lesson 1 Centimeter

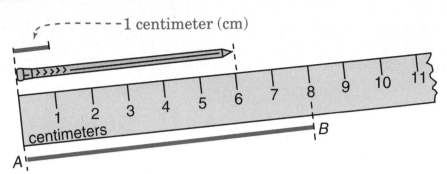

The nail is ___6___ centimeters long.   Line segment *AB* is _____ centimeters long.

Find the length of each picture to the nearest centimeter.

1. _____ cm

2. _____ cm

3. _____ cm

4. _____ cm

5. _____ cm

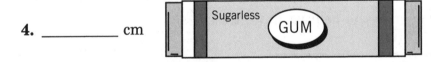

6. _____ cm

Use a ruler to draw a line segment for each measurement.

7. 5 cm

8. 8 cm

9. 3 cm

10. 6 cm

Lesson 2 Millimeter

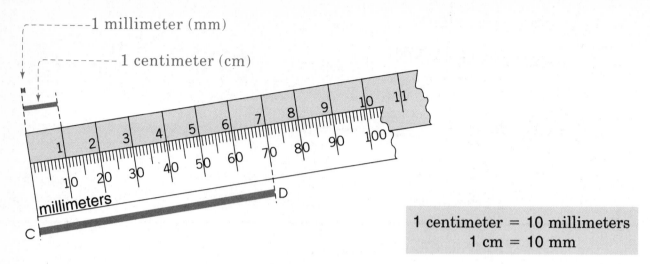

- - - - - -1 millimeter (mm)

- - - - - - 1 centimeter (cm)

| 1 centimeter = 10 millimeters |
| 1 cm = 10 mm |

Line segment *CD* is ____7____ centimeters or _____ millimeters long.

Find the length of each line segment to the nearest centimeter.
Then find the length of each line segment to the nearest millimeter.

       a   *b*

1. _____ cm   _____ mm

2. _____ cm   _____ mm

3. _____ cm   _____ mm

4. _____ cm   _____ mm

Find the length of each line segment to the nearest millimeter.

5. _____ mm

6. _____ mm

7. _____ mm

8. _____ mm

Use a ruler to draw a line segment for each measurement.

9. 50 mm

10. 80 mm

11. 25 mm

12. 55 mm

Lesson 3 Perimeter

The distance around a figure is called its perimeter.

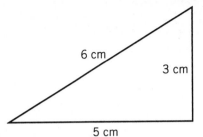

6 cm
3 cm
5 cm

$$\begin{array}{r} 6 \\ 3 \\ +5 \\ \hline 14 \end{array}$$

20 mm
20 mm
20 mm
20 mm

$$\begin{array}{r} 20 \\ 20 \\ 20 \\ +20 \\ \hline 80 \end{array}$$

The perimeter is ___14___ centimeters.   The perimeter is _____ millimeters.

Find the perimeter of each figure.

a	*b*	*c*

1.

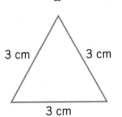

3 cm   3 cm

3 cm

20 mm

10 mm   10 mm

20 mm

15 mm

15 mm   15 mm

15 mm

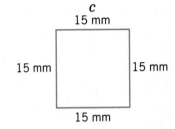

perimeter: _____ cm   *perimeter:* _____ mm   *perimeter:* _____ mm

Find the length of each side in centimeters. Then find the perimeter.

a   *b*

2.

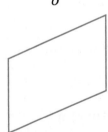

perimeter: _____ cm   *perimeter:* _____ cm

Find the length of each side in millimeters. Then find the perimeter.

3.

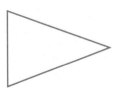

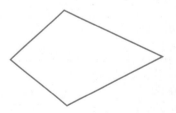

perimeter: _____ mm   *perimeter:* _____ mm

CHAPTER 14

Lesson 3 Problem Solving

Solve each problem.

1. Find the perimeter of the rectangle to the nearest centimeter.

The perimeter is _____ centimeters.

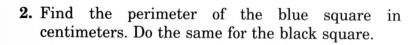

2. Find the perimeter of the blue square in centimeters. Do the same for the black square.

_____ centimeters is the perimeter of the blue square.

_____ centimeters is the perimeter of the black square.

3. How much greater is the perimeter of the blue square than that of the black square?

The perimeter is _____ centimeters greater.

What is the combined distance around the two squares?

The combined distance is _____ centimeters.

Estimate the perimeter of each of the following in centimeters. Then find each perimeter to the nearest centimeter.

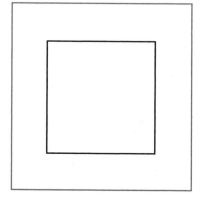

	Object	Estimate	Perimeter
4.	rectangle A	_____ cm	_____ cm
5.	triangle B	_____ cm	_____ cm
6.	cover of this book	_____ cm	_____ cm
7.	cover of a dictionary	_____ cm	_____ cm
8.	top of a chalk box	_____ cm	_____ cm

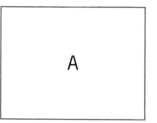

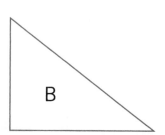

Lesson 4 Area

To find the **area** of a square or rectangle, multiply the length by the width.

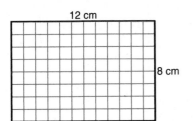

The length of the rectangle is 12 centimeters.

The width of the rectangle is 8 centimeters.

$$\begin{array}{r} 12 \\ \times\ 8 \\ \hline 96 \end{array}$$

The area of the rectangle is 96 square centimeters.

Find the area of each square or rectangle.

	a	*b*	*c*

1.

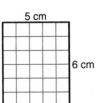

_____ square cm

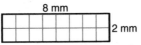

_____ square mm

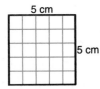

_____ square cm

2.

16 mm
4 mm

_____ square mm

15 cm
8 cm

_____ square cm

13 mm
6 mm

_____ square mm

3.

8 cm
8 cm

_____ square cm

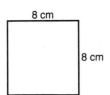
20 mm
3 mm

_____ square mm

7 cm
5 cm

_____ square cm

4.

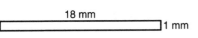
18 mm
1 mm

_____ square mm

7 cm
7 cm

_____ square cm

12 mm
15 mm

_____ square mm

Lesson 4 Problem Solving

Solve each problem.

1. The top of a rectangular table measures
   150 centimeters in length and 120 centimeters in
   width. What is the area of the table top?

   The area of the table top is _____ square cm.

2. The top of a CD case measures 12 centimeters in
   length and 11 centimeters in width. What is the
   area of the top of the CD case?

   The area of the CD case is _____ square cm.

3. A regular piece of paper measures 28 centimeters
   in length and 22 centimeters in width. What is the
   area of the paper?

   The area of the paper is _____ square cm.

4. Mrs. Schultz wears a square pin that measures
   9 millimeters on each side. What is the area of
   the pin?

   The area of the pin is _____ square mm.

5. Mr. Jefferson's car leaked oil that made a
   rectangular puddle. The puddle was 58 centimeters
   long and 39 centimeters wide. What is the area
   of the puddle?

   The area of the puddle is _____ square cm.

6. The Smith family's lawn measures 15 meters wide
   and 7 meters long. What is the area of the Smiths'
   lawn?

   The area of the lawn is _____ square meters.

1.

2.

3.

4.

5.

6.

Lesson 5 Meter and Kilometer

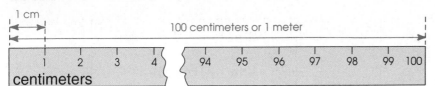

	100 centimeters = 1 meter
	100 cm = 1 m

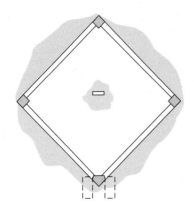

Suppose you run around a baseball diamond nine times. You would run about 1 kilometer (km).

1,000 meters = 1 kilometer
1,000 m = 1 km

Find the length of each of the following to the nearest meter.

	Object	Length
1.	width of a door	_____ m
2.	height of a door	_____ m
3.	length of a chalkboard	_____ m
4.	height of a cabinet	_____ m

Solve each problem.

5. You and some classmates lay five of your math books like this. Find the length to the nearest meter.

The length is _____ meter(s).

6. Marcus lives 3 kilometers from school. How many meters is that?

The distance is _____ meters.

7. Ms. Kahn can drive 87 kilometers in one hour. How many kilometers can she drive in four hours?

She can drive _____ kilometers in four hours.

6.

7.

CHAPTER 14

Lesson 6 Units of Length

25 cm = ___?___ mm	18 m = ___?___ cm
1 cm = 10 mm	1 m = 100 cm
1 10	1 100
×25 ×25	×18 ×18
25 250	18 1800
25 cm = _250_ mm	18 m = _1,800_ cm

9 m = ___?___ mm	7 km = ___?___ m
1 m = 1,000 mm	1 km = 1,000 m
1 1000	1 1000
×9 ×9	×7 ×7
9 9000	7 7000
9 m = _____ mm	7 km = _____ m

Complete the following.

          a   *b*

1. 9 cm = _____ mm   7 cm = _____ mm

2. 9 m = _____ cm   6 m = _____ cm

3. 9 m = _____ mm   4 m = _____ mm

4. 9 km = _____ m   5 km = _____ m

5. 16 m = _____ cm   8 m = _____ mm

6. 89 km = _____ m   46 m = _____ cm

7. 28 cm = _____ mm   18 km = _____ m

8. 13 m = _____ mm   42 cm = _____ mm

9. 16 m = _____ mm   10 m = _____ cm

10. 10 km = _____ m   25 m = _____ mm

Lesson 7 Liter and Milliliter

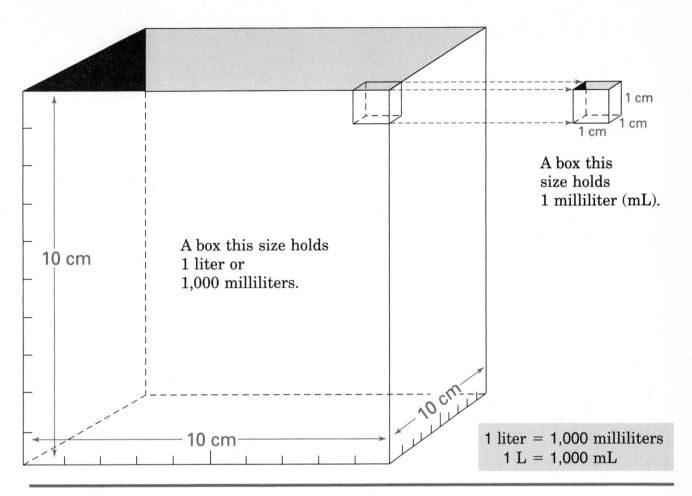

10 cm

A box this size holds
1 liter or
1,000 milliliters.

A box this
size holds
1 milliliter (mL).

1 cm
1 cm
1 cm

10 cm

10 cm

1 liter = 1,000 milliliters
1 L = 1,000 mL

Solve each problem.

1. How many milliliters does the carton hold?

The carton holds _____ milliliters.

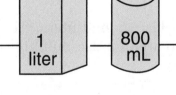

1 liter   800 mL

2. Which container holds more liquid, the carton or the bottle?

The _____ holds more liquid.

3. A small milk carton holds 236 milliliters of milk. How many milliliters of milk are in eight cartons?

There are _____ milliliters of milk.

3.

4.

4. Lyle can drive his car 11 kilometers per liter of gasoline. How far can he drive on 80 liters?

He can go _____ kilometers.

CHAPTER 14

Lesson 8 Liter and Milliliter

6 liters = ___?___ mL	13 liters = ___?___ mL
1 liter = 1,000 mL	1 liter = 1,000 mL
↓ ↓	↓ ↓
1 1000	1 1000
×6 ×6	×13 ×13
‾6‾ ‾6000‾		
↓	↓ ↓
6 liters = __6,000__ mL	13 liters = _____ mL

Complete the following.

	a		b
1.	1 liter = _____ mL		2 liters = _____ mL
2.	3 liters = _____ mL		6 liters = _____ mL
3.	7 liters = _____ mL		9 liters = _____ mL
4.	10 liters = _____ mL		11 liters = _____ mL
5.	25 liters = _____ mL		30 liters = _____ mL

Solve each problem.

6. One tablespoon holds 15 milliliters. How many tablespoons of soup are in a 225-milliliter can?

There are _____ tablespoons of soup.

6.

7. Barb used 8 liters of water when she washed her hands and face. How many milliliters of water did she use?

She used _____ milliliters of water.

There are 28 students in Barb's class. Suppose each student uses as much water as Barb. How many liters would be used?

_____ liters would be used.

7.

Lesson 9 Gram, Milligram, and Kilogram

A vitamin tablet
weighs about
100 milligrams (mg).

A dime weighs
about 2 grams.

3 of your math
books weigh about
1 kilogram (kg).

1 gram = 1,000 milligrams	1,000 grams = 1 kilogram
1 g = 1,000 mg	1,000 g = 1 kg

Use the above diagrams to do problems **1–5**.

1. What is the weight of two vitamin tablets in milligrams?

The weight is _____ milligrams.

2. Find the weight of ten vitamin tablets in milligrams. Then find the weight in grams.

The weight is _____ milligrams or ____ gram.

3. Find the weight in grams of a roll of 50 dimes.

The weight is _____ grams.

4. What is the weight in grams of ten rolls of dimes (500 dimes)? What is the weight in kilograms?

The weight is _____ grams or _____ kilogram.

5. Find the weight in kilograms of a shipment of 30 math books. Then find the weight in grams.

It is _____ kilograms or _____ grams.

1.

2.

3.

4.

5.

Tell whether you would use *milligrams, grams,* or *kilograms* to weigh each object.

a	*b*	*c*

6. a nickel _____   a grain of sand _____   a bicycle _____

7. a pin _____   a new pencil _____   yourself _____

Lesson 10 Problem Solving

Solve each problem.

1. LuKeesha found a butterfly whose wings were 6 centimeters long. How long are the butterfly's wings in millimeters?

   The butterfly's wings are_____ mm long.

   1.

2. Trevor wants to put a fence around his garden. How much fence does he need?

   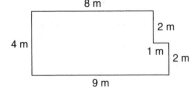

   Trevor needs _____ m of fence.

   2.

3. One tablespoon holds 15 milliliters. Rashan had to take 4 tablespoons of medicine. How many milliliters of medicine did he take?

   Rashan took _____ mL of medicine.

   3.

4. Linda can weight lift 25 kilograms at the gym. How many grams can she lift?

   Linda can lift_____ grams.

   4.

5. Sally's kitchen window measures 115 centimeters by 73 centimeters. What is the area of her window?

   The area of the window is_____ square cm.

   5.

Lesson 11   Weight

7 g = ___?___ mg
1 g = 1,000 mg
1   1000
×7   ×7
7   7000
7 g = __7,000__ mg

18 kg = ___?___ g
1 kg = 1,000 g
1   1000
×18   ×18
18   18000
18 kg = _____ g

Complete the following.

      a   *b*

1. 5 kg = _____ g   9 g = _____ mg

2. 25 g = _____ mg   78 kg = _____ g

3. Tell the weight shown on each scale.

Al's Weight Last Year   Al's Weight This Year

4. How many kilograms did Al gain?

_____ kilograms   _____ kilograms   Al gained _____ kilograms.

5. Complete the table.

Sarah's Breakfast

Food	Protein	Calcuim
1 biscuit of shredded wheat	2 g	11 mg
1 cup of whole milk	8 g	291 mg
1 banana	1 g	10 mg
Total	____ g	____ mg

6. Give the amount of protein Sarah had for breakfast in *milligrams*.

Sarah had _____ milligrams of protein.

7. How many milligrams of calcium are in 4 cups of whole milk?

_____ milligrams of calcium are in 4 cups of whole milk.

Lesson 12 Temperature (Celsius)

Use **degrees Celsius** to measure the temperature in metric units.
Read the top of the liquid in the thermometer to tell the temperature.
Write 23°C.

In degrees Celsius, water freezes at 0°C and water boils at 100°C.
In degrees Celsius, a person's normal body temperature is 37°C.
Use a minus sign to show temperatures colder than 0°C: −12°C.

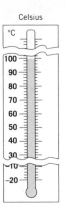

Read each thermometer and write the temperature in degrees Celsius.

| *a* | *b* | *c* | *d* |

1.

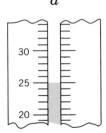   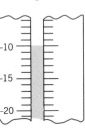   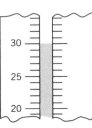   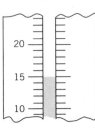

_____   _____   _____   _____

2.

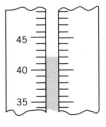   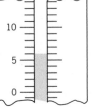   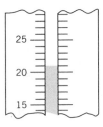

_____   _____   _____   _____

Write the best estimate of temperature for each item.

3.   ice cube   hot summer day   cup of hot cocoa   cool fall day

       −2°C or 5°C   10°C or 29°C   18°C or 75°C   16°C or 36°C

_____   _____   _____   _____

CHAPTER 14 PRACTICE TEST
Metric Measurement

Find the length of each line segment to the nearest centimeter (cm).
Then find the length of each line segment to the nearest millimeter (mm).

a	*b*	
1. _____ cm	_____ mm	▬▬▬▬▬▬▬
2. _____ cm	_____ mm	▬▬▬▬▬▬▬▬▬

Find the perimeter of each figure.

3.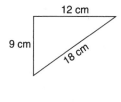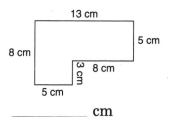

_____ cm   _____ cm   _____ cm

Find the area of each rectangle.

4.

_____ square cm   _____ square cm   _____ square cm

Write each temperature in degrees Celsius.

5.

_____ °C   _____ °C   _____ °C

Complete the following.

a	*b*
6. 5 centimeters = _____ millimeters	3 kiloliters = _____ liters
7. 7 meters = _____ centimeters	9 meters = _____ millimeters
8. 1 gram = _____ milligrams	50 centimeters = _____ millimeters
9. 5 kilograms = _____ grams	37 liters = _____ kiloliters

CHAPTER 15 PRETEST
Customary Measurement

1. Find the length of the line segment to the nearest $\frac{1}{2}$ inch.

_____ inches ▬▬▬▬▬▬▬▬▬▬▬▬▬▬▬

2. Find the length of the line segment to the nearest $\frac{1}{4}$ inch.

_____ inches ▬▬▬▬▬▬▬▬▬▬▬

Complete the following.

	a	*b*
3.	3 pounds = _____ ounces	3 tons = _____ pounds
4.	3 minutes = _____ seconds	2 hours = _____ minutes
5.	4 days = _____ hours	6 yards = _____ feet
6.	3 yards = _____ feet	4 yards = _____ inches
7.	6 feet = _____ inches	6 feet = _____ yards
8.	8 quarts = _____ pints	4 gallons = _____ quarts
9.	6 pints = _____ cups	8 pints = _____ quarts
10.	16 quarts = _____ gallons	10 cups = _____ pints

Find the perimeter of each figure.

a

11.

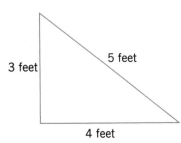

_____ feet

b

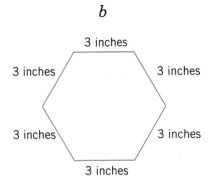

_____ inches

Lesson 1 $\frac{1}{2}$ Inch

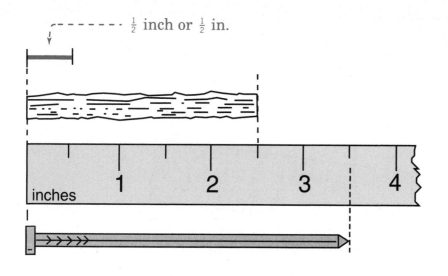

--- $\frac{1}{2}$ inch or $\frac{1}{2}$ in.

The stick is $2\frac{1}{2}$ inches long.

$2\frac{1}{2}$ is read

two and one half.

The nail is _____ inches long.

Find the length of each picture to the nearest $\frac{1}{2}$ inch.

1. _____ in.

2. _____ in.

3. _____ in.

4. _____ in.

5. _____ in. _____

6. _____ in. _____

Use a ruler to draw a line segment for each measurement.

7. $1\frac{1}{2}$ in.

8. $3\frac{1}{2}$ in.

9. $4\frac{1}{2}$ in.

10. 5 in.

CHAPTER 15

Lesson 2 $\frac{1}{4}$ Inch

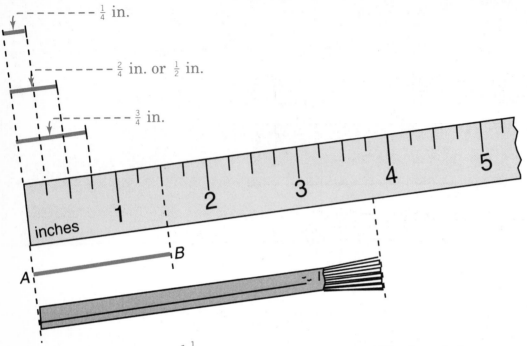

$\frac{1}{4}$ in.

$\frac{2}{4}$ in. or $\frac{1}{2}$ in.

$\frac{3}{4}$ in.

Line segment AB is __$1\frac{1}{2}$__ inches long.   The brush is _____ inches long.

Find the length of each picture to the nearest $\frac{1}{4}$ inch.

1. _____ in.

2. _____ in.

3. _____ in.

4. _____ in.

5. _____ in.

Use a ruler to draw a line segment for each measurement.

6. $2\frac{1}{4}$ in.

7. $1\frac{1}{2}$ in.

8. $4\frac{3}{4}$ in.

Lesson 3 Units of Length

6 ft = ____?____ yd

3 ft = 1 yd

$$3\overline{)6}^{\,2}$$

6 ft = ____2____ yd

1 foot (ft) = 12 in.

1 yard (yd) = 3 ft or 36 in.

1 mile (mi) = 5,280 ft

4 ft = ____?____ in.

1 ft = 12 in.

1   12

×4   ×4

4   48

4 ft = _____ in.

Complete the following.

a

b

1. 3 ft = _____ in. 12 ft = _____ yd

2. 2 yd = _____ in. 5 yd = _____ in.

3. 5 ft = _____ in. 7 ft = _____ in.

4. 2 mi = _____ ft 6 yd = _____ in.

5. 15 ft = _____ yd 27 ft = _____ yd

6. 7 yd = _____ ft 9 yd = _____ ft

7. 9 ft = _____ in. 5 mi = _____ ft

8. 15 yd = _____ ft 75 ft = _____ yd

9. 60 ft = _____ yd 7 yd = _____ in.

10. 3 yd = _____ in. 300 ft = _____ yd

11. 8 ft = _____ in. 9 yd = _____ in.

12. 10 yd = _____ ft 3 mi = _____ ft

CHAPTER 15

Lesson 3   Problem Solving

Solve each problem.

1. Mr. Jefferson is 6 feet tall. What is his height in inches?

   His height is _____ inches.

2. In baseball the distance between home plate and first base is 90 feet. What is this distance in yards?

   The distance is _____ yards.

3. Jeromy has 150 yards of kite string. How many feet of kite string does he have?

   He has _____ feet of kite string.

4. A trench is 2 yards deep. What is the depth of the trench in inches?

   The trench is _____ inches deep.

5. There are 5,280 feet in a mile. How many yards are there in a mile?

   There are _____ yards in a mile.

6. One of the pro quarterbacks can throw a football 60 yards. How many feet can he throw the football?

   He can throw the football _____ feet.

7. Marcena has 8 feet of ribbon. How many inches of ribbon does she have?

   She has _____ inches of ribbon.

8. A rope is 3 yards long. What is the length of the rope in inches?

   The rope is _____ inches long.

9. A certain car is 6 feet wide. What is the width of the car in inches?

   The car is _____ inches wide.

1.	
2.	3.
4.	5.
6.	7.
8.	9.

Lesson 4 Perimeter

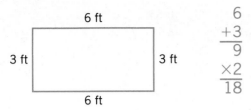

The perimeter is ___18___ feet.

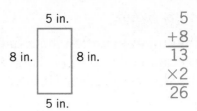

The perimeter is _____ inches.

To find the perimeter of a rectangle,
(1) add the measures of the length and width and
(2) multiply that sum by 2.

Find the perimeter of each rectangle below.

a	b	c

1.

 (15 ft, 9 ft, 9 ft, 15 ft)

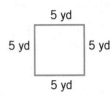 (4 in., 9 in., 9 in., 4 in.)

(5 yd, 5 yd, 5 yd, 5 yd)

_____ feet   _____ inches   _____ yards

2.

 (6 in., 9 in., 9 in., 6 in.)

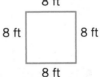 (8 ft, 8 ft, 8 ft, 8 ft)

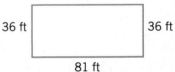 (81 ft, 36 ft, 36 ft, 81 ft)

_____ inches   _____ feet   _____ feet

3.

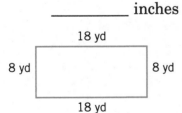 (18 yd, 8 yd, 8 yd, 18 yd)

 (12 in., 12 in., 12 in., 12 in.)

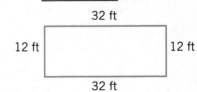 (32 ft, 12 ft, 12 ft, 32 ft)

_____ yards   _____ inches   _____ feet

4.

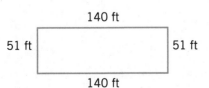 (140 ft, 51 ft, 51 ft, 140 ft)

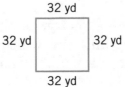 (32 yd, 32 yd, 32 yd, 32 yd)

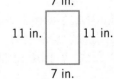 (7 in., 11 in., 11 in., 7 in.)

_____ feet   _____ yards   _____ inches

Lesson 4   Problem Solving

Solve each problem.

1. Mr. Champney's garden is shaped like a rectangle. The rectangle is 24 feet long and 16 feet wide. What is the perimeter of his garden?

   The perimeter of his garden is _____ feet.

2. A rectangular desktop is 24 inches long and 16 inches wide. What is the perimeter of the desktop?

   The perimeter of the desktop is _____ inches.

3. A flower garden is shaped like a rectangle. The length of the rectangle is 40 feet and the width is 30 feet. How many feet of edging will be needed to go around the garden?

   _____ feet of edging will be needed.

4. A rectangular windowpane is 28 inches long and 24 inches wide. What is the perimeter of the windowpane?

   The perimeter is _____ inches.

5. Mrs. Richardson has a rectangular-shaped mirror that is 4 feet long and 3 feet wide. How many feet of ribbon will she need to go around the edges of the mirror?

   She will need _____ feet of ribbon.

6. A rectangular picture frame is 32 inches long and 24 inches wide. What is the perimeter of the picture frame?

   The perimeter is _____ inches.

7. A football field is shaped like a rectangle. The length of the field is 360 feet and the width is 160 feet. What is the perimeter of a football field?

   The perimeter is _____ feet.

1.	
2.	**3.**
4.	**5.**
6.	**7.**

Lesson 5 Area

To find the **area** of a square or rectangle, multiply the length by the width.

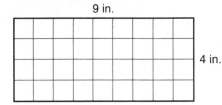

The length of the rectangle is 9 inches.
The width of the rectangle is 4 inches.

$$\begin{array}{r} 9 \\ \times\, 4 \\ \hline 36 \end{array}$$

The area of the rectangle is 36 square inches.

Find the area of each square or rectangle.

a	*b*	*c*

1.

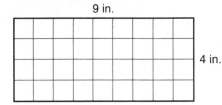

_____ square in.   _____ square ft   _____ square in.

2.

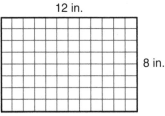 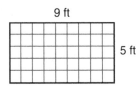 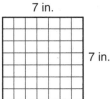

_____ square ft   _____ square in.   _____ square ft

3.

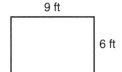 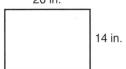

_____ square in.   _____ square ft   _____ square in.

4.

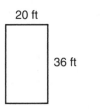 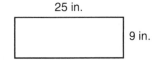

_____ square ft   _____ square in.   _____ square ft

CHAPTER 15

Lesson 5 Problem Solving

Solve each problem.

1. Derek painted a wall that was 15 feet long and 8 feet high. What area did he paint?

   The area of the wall is _____ square ft.

2. Jackie made a bed for her dog that was 24 inches wide and 42 inches long. What is the area of the dog bed?

   The area of the dog bed is _____ square in.

3. A computer screen measures 12 inches high and 14 inches wide. What is the area of the computer screen?

   The area of the screen is _____ square in.

4. Isaac put a new rug in his living room. The rug measures 18 feet long and 15 feet wide. What is the area of the rug?

   The area of the rug is _____ square ft.

5. Mr. and Mrs. Kovac plan to build a new house that will be 52 feet long and 38 feet wide. How much area will the house cover?

   The area of the house is _____ square ft.

6. Kendra painted a picture that was 36 inches wide and 21 inches tall. What is the area of the painting?

   The area of the painting is _____ square in.

1.

2.

3.

4.

5.

6.

Lesson 6 Capacity

$$6 \text{ qt} = \underline{\quad ? \quad} \text{ pt}$$

$$
\begin{array}{cc}
1 \text{ qt} = & 2 \text{ pt} \\
\downarrow & \downarrow \\
1 & 2 \\
\times 6 & \times 6 \\
\hline
6 & 12 \\
\downarrow & \searrow \\
\end{array}
$$

$$6 \text{ qt} = \underline{\qquad} \text{ pt}$$

1 pint (pt) = 2 cups

1 quart (qt) = 2 pt

1 gallon (gal) = 4 qt

$$12 \text{ qt} = \underline{\quad ? \quad} \text{ gal}$$

$$4 \text{ qt} = 1 \text{ gal}$$

$$4\overline{)1\,2}^{\;3}$$

$$12 \text{ qt} = \underline{\qquad} \text{ gal}$$

Complete the following.

	a	b
1.	6 cups = _____ pt	12 qt = _____ pt
2.	4 pt = _____ qt	8 gal = _____ qt
3.	8 qt = _____ gal	6 pt = _____ cups
4.	8 pt = _____ cups	24 qt = _____ gal
5.	10 qt = _____ pt	18 pt = _____ qt
6.	5 gal = _____ qt	10 cups = _____ pt
7.	10 gal = _____ qt	32 qt = _____ gal
8.	12 pt = _____ qt	18 cups = _____ pt
9.	10 pt = _____ cups	32 pt = _____ qt
10.	28 qt = _____ gal	12 gal = _____ qt
11.	16 qt = _____ pt	28 qt = _____ pt
12.	8 cups = _____ pt	16 pt = _____ cups

Lesson 6   Problem Solving

Solve each problem.

1. There are 6 pints of lemonade in a picnic cooler. How many 1-cup containers can be filled by using the lemonade in the cooler?

    _____ containers can be filled.

2. The cooling system on a car holds 16 quarts. How many gallons does it hold?

    It holds _____ gallons.

3. In problem **2,** how many pints does the cooling system hold?

    It holds _____ pints.

4. There are 376 quarts of milk delivered to the store. How many gallons of milk is this?

    It is _____ gallons of milk.

5. How many quarts of water would be needed to fill a 10-gallon aquarium?

    _____ quarts would be needed.

6. The lunchroom served 168 pints of milk at lunch. How many quarts of milk was this?

    It was _____ quarts of milk.

7. There are 12 cups of liquid in a container. How many 1-pint jars can be filled by using the liquid in the container?

    _____ jars can be filled.

8. There are 6 pints of bleach in a container. How many quarts of bleach are in the container?

    There are _____ quarts of bleach in the container.

1.	2.
3.	4.
5.	6.
7.	8.

Lesson 7 Weight

Ounce, pound, and **ton** are customary units of weight.

5 lb = ____?____ oz
1 lb = 16 oz
↓   ↓
1   16
×5   ×5
5   80
↓   ↓
5 lb = __80__ oz

2 T = ____?____ lb
1 T = 2,000 lb
↓   ↓
1   2000
×2   ×2
2   4000
↓   ↓
2 T = _____ lb

> 1 pound (lb) = 16 ounces (oz)
> 1 ton (T) = 2,000 lb

Complete the following.

a   *b*

1. 2 lb = _____ oz   6 T = _____ lb

2. 2 T = _____ lb   4 lb = _____ oz

3. 7 lb = _____ oz   5 T = _____ lb

4. Tell the weight shown on each scale.

Tom's weight last year

KILOGRAMS ▼
60   70

_____ pounds

Tom's weight this year

KILOGRAMS ▼
60   70   80

_____ pounds

5. How many pounds did Tom gain?

Tom gained _____ pounds since last year.

Complete the table.

	Tons	Pounds	Ounces
6.		2,000	
7.	4		
8.	7		224,000

Lesson 7 Problem Solving

Solve each problem.

1. Mrs. Wilson bought a 12-pound turkey. How many ounces did the turkey weigh?

   The turkey weighed _____ oz.

   1.

2. The Garden Club grew 2 tons of watermelons to sell. How many pounds of watermelons did they grow?

   The club grew _____ pounds of watermelons.

   2.

3. Mason bought 50 pounds of apples to make applesauce. How many ounces of apples did he buy?

   Mason bought _____ ounces of apples.

   3.

4. Brett bought a truck that can hold 25 tons of stone. How many pounds of stone could the truck hold?

   The truck can hold _____ pounds of stone.

   4.

5. Jack and Beth picked 163 pounds of blueberries. How many ounces of blueberries did they pick?

   They picked _____ ounces of blueberries.

   5.

6. An African elephant can weigh 6 tons. How many pounds can an African elephant weigh?

   An African elephant can weigh _____ pounds.

   6.

Lesson 8 Time

Second, minute, hour, and **day** are units of time.

1 minute (min) = 60 seconds (sec)

1 hour = 60 min

1 day = 24 hours

3 min = ___?___ sec

1 min = 60 sec

1	60
×3	×3
3	180

3 min = _____ sec

Complete the following.

a   *b*

1. 2 hours = _____ min   8 min = _____ sec

2. 2 days = _____ hours   5 hours = _____ min

3. 5 min = _____ sec   3 days = _____ hours

4. 12 hours = _____ min   6 min = _____ sec

5. 5 days = _____ hours   10 min = _____ sec

6. 24 hours = _____ min   7 days = _____ hours

7. 4 min = _____ sec   8 hours = _____ min

8. 13 min = _____ sec   10 days = _____ hours

9. 6 hours = _____ min   4 days = _____ hours

10. 15 min = _____ sec   3 hours = _____ min

11. 14 days = _____ hours   7 min = _____ sec

12. 4 hours = _____ min   8 days = _____ hours

13. 30 min = _____ sec   48 hours = _____ min

14. 30 days = _____ hours   25 min = _____ sec

15. 15 hours = _____ min   9 days = _____ hours

CHAPTER 15

Lesson 9 Temperature (Fahrenheit)

Use **degrees Fahrenheit** to measure the temperature in customary units. Read the top of the liquid in the thermometer to tell the temperature. Write 77°F.

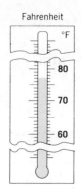

In degrees Fahrenheit, water freezes at 32°F.

In degrees Fahrenheit, water boils at 212°F.

Your normal body temperature is about 99°F.

Record the temperature shown on each thermometer.

1.

 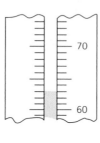 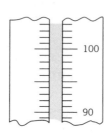 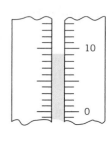

_____°F   _____°F   _____°F   _____°F

2. Water freezes at 32°F and boils at 212°F. What is the difference between those two temperatures?

The difference is _____ degrees.

2.

3. At 6 A.M. the temperature was 56°F. The high temperature was expected to be 35 degrees warmer than that. What was the high temperature expected to be?

_____°F was the expected high temperature.

3.

4. During a windy day, the windchill was 6°F. With no wind, the temperature would have been 29 degrees warmer. What would have been the temperature with no wind?

The temperature would have been _____°F.

4.

Lesson 10   Problem Solving

Solve each problem.

1. David wants to put a fence around his pool for safety reasons. The pool is 32 feet long and 20 feet wide. How much fencing does David need?

David needs _____ feet of fencing.

2. David also wants to get a cover for his pool. How many square feet will he need to cover?

David needs to cover _____ square feet.

3. Sonia had to wait three hours in the doctor's office. How many minutes did she have to wait?

Sonia had to wait _____ minutes.

4. There are 8 pints of water in a container. How many 1-cup canteens can be filled by using the water in the container?

_____ canteens can be filled.

5. A football field is 100 yards long. What is that distance in feet?

The football field is _____ feet long.

6. The American bison can live 33 years. How many days can an American bison live?

An American bison can live _____ days.

1.

2.

3.

4.

5.

6.

CHAPTER 15

Lesson 10 Problem Solving

Solve each problem.

7. A building on Main Street is 128 feet tall. What is its height in inches?

   The building is _____ inches tall.

7.

8. A board is 144 feet long. What is the length of the board in yards?

   The board is _____ yards long.

8.

9. A rectangular picture frame is 28 inches wide and 36 inches long. What is the perimeter of the frame?

   The perimeter of the frame is _____ inches.

9.

10. The side of a barn is 16 feet high and 120 feet long. What is the area of the side of the barn?

    The area of the side of the barn is _____ square feet.

10.

11. How many quarts of soup would be needed to fill a 3-gallon pot?

    _____ quarts would be needed.

11.

12. According to the story, Rip Van Winkle slept for 20 years. A year has 365 days. How many days did he sleep?

    Rip Van Winkle slept for _____ days.

12.

13. The Observation Platform on the 102nd floor of the Empire State Building in New York City is 1,224 feet high. How many yards tall is the Observation Platform?

    The Observation Platform on the 102nd floor is _____ yards tall.

13.

CHAPTER 15 PRACTICE TEST
Customary Measurement

Find the length of each line segment to the nearest $\frac{1}{4}$ inch.

a	b

1. ▬▬▬▬▬▬▬   ▬▬▬▬▬▬▬▬▬▬

_____ in.   _____ in.

Find the perimeter of each figure.

| a | b | c |

2.

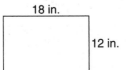

18 in. / 12 in.

36 in. / 5 in.

9 in. / 9 in.

_____ in.   _____ in.   _____ in.

Find the area of each quadrilateral.

3.

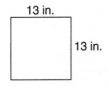

13 in. / 13 in.

13 ft / 22 ft

9 in. / 16 in.

_____ square in.   _____ square ft   _____ square in.

Complete the following.

| a | b |

4. 5 feet = _____ inches   3 hours = _____ minutes

5. 7 quarts = _____ pints   9 yards = _____ feet

6. 5 pounds = _____ ounces   15 minutes = _____ seconds

CHAPTER 15 PRACTICE TEST

CHAPTER 16 PRETEST
Graphs and Probability

Use the bar graph to answer each question.

1. How many snow shovels did the store sell in November?

The store sold _____ snow shovels.

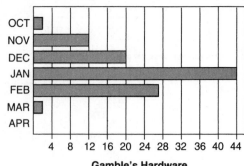

Gamble's Hardware
Number of Snow Shovels Sold

2. How many more snow shovels were sold in January than in March?

_____ more snow shovels were sold in January.

Use the line graph to answer each question.

3. How many bats were sighted at 12 A.M.?

_____ bats were sighted.

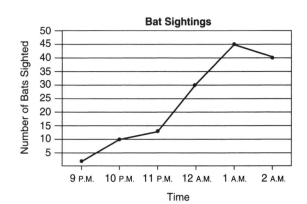

4. How many more bats were sighted at 1 A.M. than at 10 P.M.?

_____ more bats were sighted.

Find the median, mode, and range of each set of numbers.

5. 6, 8, 3, 2, 9, 4, 6, 11, 15, 9, 4, 2, 9

The median is _____.

The mode is _____.

The range is _____.

15, 12, 17, 15, 26, 17, 31, 26, 17

The median is _____.

The mode is _____.

The range is _____.

Refer to the wheel.

6. What is the probability of spinning a 4?

The probability is _____.

Lesson 1 Bar Graphs

A **bar graph** compares information. The information is shown on the graph by bars.

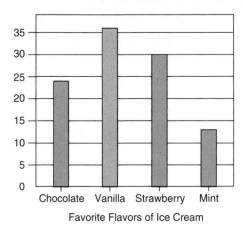

Favorite Flavors of Ice Cream

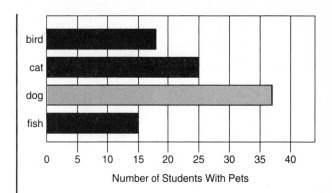

Number of Students With Pets

This graph compares favorite flavors of ice cream.

The bar for vanilla is the longest, so most people choose vanilla as their favorite flavor.

This graph compares what pets students have at home.

The bar for dogs is the longest. 37 students have pet dogs.

Read each graph and answer the questions.

1.

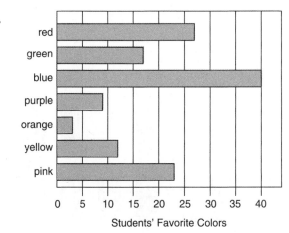

Students' Favorite Colors

What is the favorite color shown by the graph?

_____ is the favorite color.

Which color is the favorite color of 9 students?

_____ is the favorite color of 9 students.

2.

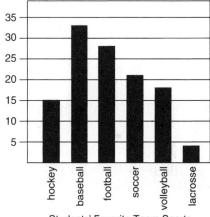

Students' Favorite Team Sports

What is the favorite game shown by the graph?

_____ is the favorite game.

What is the least favorite game?

_____ is the least favorite game.

CHAPTER 16

Lesson 1   Problem Solving

Use the bar graph to answer each question.

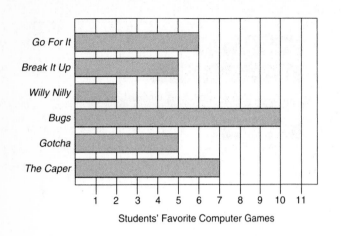

Students' Favorite Computer Games

1. What is the favorite computer game shown by the bar graph?

    _____ is the favorite computer game.

2. How many students like the game *Go For It?*

    _____ students like *Go For It*.

3. How many students like the game *Willy Nilly?*

    _____ students like *Willy Nilly*.

4. How many more students like *The Caper* than *Willy Nilly?*

    _____ more students like *The Caper*.

Use the bar graph to answer each question.

5. What creature can hold its breath the longest?

    A _____ can hold its breath the longest.

6. How long can the average platypus hold its breath?

    A platypus can hold its breath _____ minutes.

7. Which creature can hold its breath longer, a platypus or a muskrat?

    A _____ can hold its breath longer.

8. How much longer can a hippopotamus hold its breath than a sea otter?

    A hippopotamus can hold its breath _____ more minutes than a sea otter.

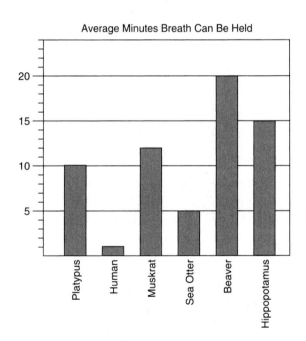

Average Minutes Breath Can Be Held

Lesson 2   Line Graphs

A **line graph** shows how something changes over time. The information is shown on the graph by a line that connects points on the graph.

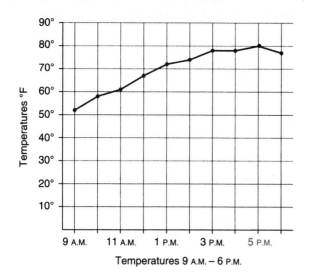

Temperatures 9 A.M. – 6 P.M.

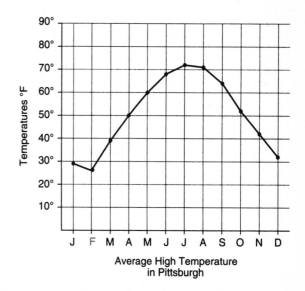

Average High Temperature
in Pittsburgh

This graph shows how the temperature changed during one day.

The point for 80°F is the highest point, so was at the highest temperature for the day was at 5 P.M.

This graph shows the average temperature in Pittsburgh for one year.

The point for February is the lowest, so that was the coldest month.

Study each graph and answer the questions.

1.

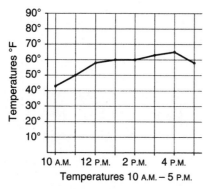

Temperatures 10 A.M. – 5 P.M.

What is the highest temperature shown by the graph?

_____ is the highest temperature.

At what time was the temperature 63°F?

_____

2.

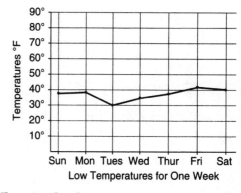

Low Temperatures for One Week

What is the lowest temperature shown by the graph?

_____ is the lowest temperature.

What day had the highest temperature?

_____ had the highest temperature.

CHAPTER 16

Lesson 2 Problem Solving

Use the line graph to answer each question.

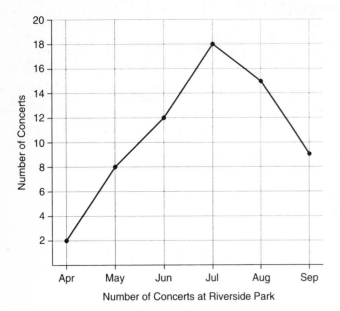

Number of Concerts at Riverside Park

1. In which month were the most concerts held at Riverside Park?

The most concerts were held in _____.

2. How many concerts were held in August?

There were _____ concerts held in August.

3. How many more concerts were held in July than in April?

There were _____ more concerts in July.

Use the line graph to answer each question.

4. On which day did John swim the most laps?

He swam the most laps on _____.

5. How many laps did John swim on Tuesday?

He swam _____ laps on Tuesday.

6. On which day did John swim the fewest laps?

He swam the fewest laps on _____.

7. How many more laps did John swim on Wednesday than he swam on Monday?

He swam _____ more laps on Wednesday.

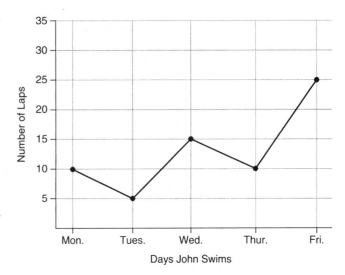

Days John Swims

Lesson 3 Mean

The **mean,** or average, of a set of numbers is the sum of the numbers divided by the number of addends.

Find the mean of this set of numbers: 6, 7, 12, 16, 24.

$$
\begin{array}{r}
6 \\
7 \\
12 \\
16 \\
+24 \\
\hline
65
\end{array}
$$

There are five addends.
Divide by 5.

$$5 \overline{)65} \quad \overset{13}{}$$

The mean of this set of numbers is _____.

Find the mean of each set of numbers.

a	*b*

1. 5, 14, 17, 18, 21   25, 26, 27, 28, 29

The mean is _____.   The mean is _____.

2. 15, 18, 19, 25, 28, 33   23, 29, 36, 38, 41, 43

The mean is _____.   The mean is _____.

3. 10, 12, 14, 16, 18, 20, 22   20, 25, 30, 35, 40, 45, 50

The mean is _____.   The mean is _____.

4. 2, 3, 5, 8, 9, 11, 12, 14   9, 13, 16, 17, 23, 25, 26, 31

The mean is _____.   The mean is _____.

5. 2, 3, 5, 5, 7, 8, 9, 9, 15   3, 3, 5, 5, 8, 9, 11, 13, 15

The mean is _____.   The mean is _____.

6. 104, 105, 112, 113, 116   132, 144, 145, 147, 150, 152

The mean is _____.   The mean is _____.

7. 204, 237, 244, 259   21, 24, 25, 27, 32, 38, 39, 42

The mean is _____.   The mean is _____.

CHAPTER 16

Lesson 3 Problem Solving

Answer each question.

1. Marcia received grades of 98, 88, 95, and 83 on her math tests. What is Marcia's mean score?

   The mean score is _____.

1.

2. The following times were turned in at the obstacle course: 45 minutes, 52 minutes, 57 minutes, 49 minutes, 41 minutes, and 56 minutes. What is the mean time for running the obstacle course?

   The mean time is _____ minutes.

2.

3. Sheldon counted the number of fish in each tank at the pet store. He counted 32, 16, 44, 27, and 46 fish. What is the mean number of fish?

   The mean number of fish is _____.

3.

4. The chart shows the temperature at several times during the day. What was the mean temperature for the day?

8 A.M.	56°F	12 P.M.	75°F
9 A.M.	58°F	1 P.M.	75°F
10 A.M.	61°F	2 P.M.	76°F
11 A.M.	63°F	3 P.M.	72°F

   The mean temperature was _____°F.

4.

5. Jamal scored 21, 35, 10, 17 and 22 points in his last five basketball games. What is the mean number of points scored in his last five games?

   The mean number of points is _____.

5.

Lesson 4   Median, Mode, and Range

The **median** of a set of numbers is the middle number when the numbers are ordered least to greatest.

$$5, 2, 8, 6, 4, 9, 2$$

$$2, 2, 4, \underline{5}, 6, 8, 9$$

The median is 5.

The **mode** of a set of numbers is the number that occurs most often.

$$5, 2, 8, 6, 4, 9, 2$$

$$\underline{2}, \underline{2}, 4, 5, 6, 8, 9$$

The mode is 2.

The **range** of a set of numbers is the difference between the greatest number and the least number.

$$5, \underline{2}, 8, 6, 4, \underline{9}, \underline{2}$$

The greatest number is 9.
The least number is 2.

$$9 - 2 = 7$$

The range is 7.

Find the median, mode, and range for each set of numbers.

	a	*b*	*c*

1. 6, 2, 8, 9, 1, 3, 2

The median is _____.   The mode is _____.   The range is _____.

2. 5, 9, 3, 1, 8, 5, 7, 4, 12

The median is _____.   The mode is _____.   The range is _____.

3. 7, 3, 2, 9, 5, 6, 3, 1, 3, 7, 5

The median is _____.   The mode is _____.   The range is _____.

4. 3, 7, 5, 1, 7, 4, 6, 3, 8, 14, 23, 1, 7, 8, 7,

The median is _____.   The mode is _____.   The range is _____.

5. 11, 6, 5, 2, 6, 13, 7, 6, 19, 8, 14, 6, 2, 4, 10

The median is _____.   The mode is _____.   The range is _____.

CHAPTER 16

Lesson 4   Problem Solving

Solve each problem.

1. The following scores were turned in by the Lions baseball team: 11, 9, 3, 0, 6, 7, 4, 9, and 5. What is the median number of runs the team scored?

   The median number of runs scored is _____.

   1.

2. The following scores were turned in by the Panthers football team: 35, 7, 6, 21, 28, 14, 17, 28, and 9. What is the range of the team's scores?

   The range of the scores is_____.

   2.

3. Dr. Berry noticed the following weights of rabbits that were brought to her office: 8 lb, 5 lb, 12 lb, 7 lb, 6 lb, 10 lb, 11 lb, 9 lb, 13 lb, and 9 lb. What is the mode of the weights?

   The mode of the weights is _____ lb.

   3.

4. The Video Shack rented the following numbers of videos last week: Mon., 352; Tues., 244; Wed., 198; Thurs., 141; Fri., 378; Sat., 831; Sun., 211. What is the median number of movies rented?

   The median number of videos rented is _____.

   4.

5. A survey found students watched the following numbers of hours of TV per week: 24, 12, 35, 28, 7, 21, 24, 31, 20, 19, and 15. What is the range of hours?

   The range of hours is _____.

   5.

Lesson 5 Probability

Use a fraction to describe the **probability** of something happening.

What is the probability of spinning a 3 on this wheel?

There are eight possible outcomes of a spin:
1, 1, 2, 2, 3, 3, 3, and 4.
Use the number of **outcomes** as the denominator of the fraction.

$$\frac{?}{8}$$

There are three ways to spin a 3. Use this number as the numerator of the fraction.

$$\frac{3}{8}$$

The probability of spinning a 3 is $\frac{3}{8}$.

Find each probability.

1. What is the probability of spinning a 6 on this wheel?

The probability of spinning a 6 is _____.

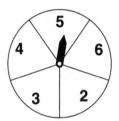

2. What is the probability of spinning a 4 on this wheel?

The probability is _____.

3. What is the probability of spinning a 2 on this wheel?

The probability is _____.

4. What is the probability of tossing a head on this coin?

The probability is _____.

CHAPTER 16

Lesson 5 Problem Solving

Find each answer.

1. Pablo has five black socks and seven white socks in a drawer. If he reaches in and takes one sock without looking, what is the probability he will get a white sock?

   The probability of getting a white sock is _____.

2. Cayce planted four red roses, eight yellow roses, and six pink roses. What is the probability that a red rose will be the first to bloom?

   The probability that a red rose blooms first is

   _____.

3. Darnell has six books about cars and eight books about spiders. If he takes one book off the shelf without looking, what is the probability he will get a book about spiders?

   The probability he will pick a book about spiders is

   _____.

4. Ali collects rocks. His collection has 15 samples of quartz, 7 samples of mica, and 12 samples of pyrite. If his cat knocks one rock off the shelf, what is the probability that it is a sample of mica?

   The probability the cat will knock off a sample of

   mica is _____.

5. Karine has a bag with 14 blue marbles, 6 purple marbles, and 5 green marbles. If she closes her eyes and picks a marble, what is the probability that she chooses a purple marble?

   The probability of picking a purple marble is

   _____.

1.

2.

3.

4.

5.

CHAPTER 16 PRACTICE TEST
Graphs and Probability

Use the bar graph to answer each question.

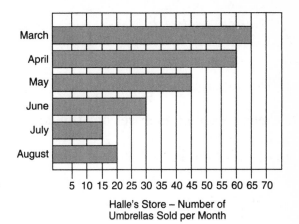

Halle's Store – Number of Umbrellas Sold per Month

1. How many umbrellas did the store sell in July?

    The store sold _____ umbrellas.

2. How many more umbrellas were sold in March than in August?

    _____ more umbrellas were sold in

Use the line graph to answer each question.

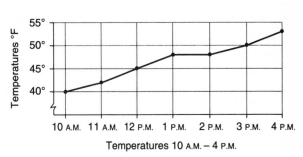

Temperatures 10 A.M. – 4 P.M.

3. What was the temperature at 12 P.M.?

    The temperature was _____ °F.

4. How much did the temperature rise between 10 A.M. and 3 P.M.?

    The temperature rose _____ degrees.

Find the mean of each set of numbers.

| a | b |

5. 2, 3, 7, 9, 13, 15, 21   37, 38, 41, 41, 41, 43, 47, 48, 51

    The mean is _____.   The mean is _____.

Find the median of each set of numbers.

6. 4, 2, 7, 9, 3, 5, 7, 1, 5   12, 15, 13, 18, 12, 19, 14, 14, 17

    The median is _____.   The median is _____.

CHAPTER 16

Find the mode of each set of numbers.

a   b

7. 3, 7, 4, 5, 9, 3, 7, 2, 1, 6, 4, 5, 7   14, 15, 17, 12, 14, 19, 15, 14, 17, 18, 11

The mode is _____.   The mode is _____.

Find the range of each set of numbers.

8. 4, 7, 6, 8, 4, 9, 5, 7, 6   67, 25, 94, 57, 25, 43, 66, 19

The range is _____.   The range is _____.

Refer to the wheel. Find each probability.

9. What is the probability of spinning a 2 on this wheel?

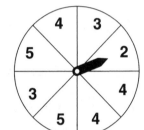

The probability is _____.

10. What is the probability of spinning a 5 on this wheel?

The probability is _____.

11. What is the probability of spinning a 4 on this wheel?

The probability is _____.

12. What is the probability of spinning a 1 on this wheel?

The probability is _____.

13. What is the probability of spinning a 3 on this wheel?

The probability is _____.

MID-TEST 1 Chapters 1-7

Add.

	a	b	c	d	e	f
1.	3 2 +4	6 1 +7	4 3 +6	7 5 +2	6 4 +4	2 3 +5
2.	8 9 +3 6	1 4 3 +5 3	5 8 +8 7	2 5 0 +9 2	3 4 7 +4 9	5 2 9 +6 7
3.	6 4 8 +3 8 2	5 9 3 +2 8 8	3 7 0 +5 2 5	4 9 7 3 +3 7 2 8	7 3 8 5 +9 2 6 4	4 8 3 7 +1 3 3 7
4.	6 3 7 9 4 +6 3 4 8	4 3 8 8 1 +5 2 3 1	3 2 6 8 4 +5 8 3 5	5 3 0 5 8 +7 2 6 9	3 2 6 6 5 +3 3 5 8	1 8 5 6 9 +7 3 4 6

Subtract.

	a	b	c	d	e	f
5.	6 8 −4	3 7 −7	5 8 −6	4 6 −2	2 7 −4	8 8 −5
6.	6 7 −2 8	2 6 4 −7 6	9 1 −4 7	3 0 6 −8 8	4 2 0 −6 5	7 2 1 −8 4
7.	5 4 7 −1 5 8	7 4 3 −1 1 9	4 7 0 −2 9 5	7 3 4 1 −3 6 9 5	5 9 9 3 −5 2 7 6	8 0 5 1 −4 8 6 2
8.	3 6 5 4 7 −7 3 4 1	7 2 3 6 1 −8 3 4 7	5 0 2 1 6 −5 7 3 4	6 6 8 3 4 −2 9 4 5	9 2 5 4 3 −4 0 2 2	2 7 4 4 5 −2 9 9 5

Solve each problem.

9. The Brooklyn Tunnel is 9,117 feet long. The Hampton Roads Tunnel is 7,479 feet long. How much longer is the Brooklyn Tunnel?

The Brooklyn Tunnel is _____ feet longer.

9.

10. It is 1,198 miles from New York to Kansas City and 1,589 miles from Kansas City to Los Angeles. How far is it from New York to Los Angeles?

It is _____ miles from New York to Los Angeles.

10.

GO

Multiply.

	a	b	c	d	e	f
11.	4 7 ×5	2 6 ×9	6 3 ×7	8 2 ×3	3 7 ×8	5 5 ×5
12.	7 3 ×1 5	2 8 ×4 3	9 3 ×3 7	4 2 6 ×9 4 5	7 3 4 ×3 3 7	9 9 2 ×4 8 3
13.	3 8 5 6 ×3 8	5 2 3 0 ×8 4	5 7 7 1 ×3 6	2 9 6 8 ×1 2	8 4 3 7 ×7 3	3 8 5 6 ×5 9

Solve each problem.

14. A marathon is about 26 miles long. Emil ran in 15 marathons this year. How far did he run?

Emil ran about _____ miles.

14.

15. The mall is 1,967 feet from one end to the other. The security guard walks the length of the mall 24 times each day. How far does he walk each day?

He walks _____ feet.

15.

Ring the correct answer.

16. Which number has a 3 in the hundredths place? $3.75 $4.83 $1.39

17. Which number has a 7 in the tenths place? $2.97 $7.44 $5.73

Add or subtract. Watch the signs.

	a	b	c	d	e	f
18.	7 5¢ −4 9¢	5 8¢ +2 7¢	9 4¢ −5 6¢	$2.5 8 −0.8 5	$1 2.6 1 +1 0.9 4	$9.4 4 +8.6 8

STOP

MID-TEST 2 Chapters 1-11

Add.

	a	*b*	*c*	*d*	*e*	*f*
1.	2 2 +9 7	3 0 4 +7 8	5 5 +8 2	5 8 3 +5 9	7 6 5 +8 4	7 9 3 +6 6
2.	4 8 7 +1 7 5	3 7 6 +6 9 3	6 9 3 4 +5 8 3 3	5 9 3 3 +6 7 4 9	7 8 6 0 4 +4 5 8 3	7 9 5 3 9 +8 9 3 0

Subtract.

3.	7 8 –3 4	4 8 7 –9 1	9 6 –2 8	3 9 7 –5 4	8 8 5 –8 6	7 9 3 –7 5
4.	7 6 9 –2 7 6	7 9 5 –3 8 8	7 1 1 –1 9 9	6 9 8 9 –5 9 2 3	7 9 5 6 –2 7 5 6	9 0 4 8 –3 7 9 0
5.	6 9 7 3 1 –4 7 2 2	6 4 4 8 7 –5 8 3 7	8 4 4 9 1 –9 4 7 3	6 8 9 4 5 –5 7 3 8	9 4 7 7 1 –8 5 6 6	6 8 4 5 6 –9 4 9 9

Solve each problem.

6. Mario has a collection of 3,587 stamps. Juan has a collection of 4,178 stamps. How many stamps do they have in all?

They have _____ stamps in all.

6.

7. How many more stamps does Juan have than Mario?

Juan has _____ more stamps.

7.

Ring the correct answer.

8. Which number has the 6 in the hundredths place? $2.46 $6.24 $4.62

9. Which number has a 0 in the tenths place? $0.47 $7.04 $4.70

GO ▶

MID-TEST 2 Chapters 1-11 (Continued)

Add or subtract. Watch the signs.

	a	*b*	*c*	*d*	*e*	*f*
10.	3 8¢ −1 6¢	3 8¢ +4 8¢	8 8¢ −3 9¢	$4.2 8 −0.7 8	$2 3.5 6 +1 2.8 7	$7.2 3 +6.9 7

Multiply.

11.	3 6 ×1 2	7 4 ×2 8	7 7 ×4 5	9 1 ×7 2	5 6 ×1 8	6 9 ×4 1
12.	1 0 8 ×5 8	5 8 4 ×3 5	6 8 4 ×8 2	5 0 9 ×3 3 4	7 1 9 ×4 2 8	7 9 5 ×6 6 3
13.	7 3 5 2 ×2 3	4 8 8 2 ×6 7	9 4 5 0 ×8 3	5 2 7 6 ×9 7	8 3 2 1 ×4 6	3 9 9 5 ×3 8

Divide.

14.	4)1 6	2)1 8	6)3 6	8)5 6	7)3 5	9)6 3
15.	5)3 4	6)5 2	4)3 7	9)2 2	8)4 0	6)3 9
16.	3)5 7	2)4 5	6)8 2	3)7 7	8)9 2	5)6 7

Estimate each answer.

	a	*b*	*c*	*d*	*e*
17.	5)3 7 5	6)2 8 8	8)4 0 6	7)8 4 3	4)8 4 9
18.	5 8 7 6 ×8	3 4 6 2 ×2	1 6 7 4 ×8	8 3 7 ×4	2 8 7 6 ×3

STOP

FINAL TEST Chapters 1-16

Add.

	a	*b*	*c*	*d*	*e*
1.	58 +21	168 +95	27 +49	308 +84	292 +75
2.	386 +488	559 +482	602 +594	5835 +4827	6482 +8273
3.	19875 +6582	38549 +3885	95763 +2994	92807 3967 +5381	39581 5387 476 +1883

Subtract.

	a	*b*	*c*	*d*	*e*
4.	85 -3	59 -6	38 -6	44 -3	69 -7
5.	78 -38	389 -66	84 -29	245 -56	800 -73
6.	970 -476	843 -328	785 -444	7289 -2846	7531 -7058

Add or subtract.

	a	*b*	*c*	*d*	*e*
7.	528 +923	5669 -736	8468 -4372	5723 +1087	3578 +2889

GO →

FINAL TEST
Chapters 1-16

Multiply.

	a	*b*	*c*	*d*	*e*
8.	58 ×5	37 ×9	82 ×7	41 ×3	95 ×8
9.	65 ×21	87 ×34	28 ×73	45 ×54	82 ×92
10.	584 ×26	209 ×75	759 ×48	914 ×264	688 ×237
11.	1875 × 26	3880 × 70	6819 × 23	4938 × 59	7592 × 86

Divide.

12. 5)25   3)18   7)28   6)43   4)39

13. 3)82   2)51   6)72   3)85   8)77

14. 5)835   6)288   8)804   7)633   4)189

15. 7)2864   5)2668   3)7658   8)9728   7)3547

GO

FINAL TEST Chapters 1-16 (continued)

Multiply or divide. Check each answer.

	a	b	c
16.	3 2 6 × 7	8) 1 4 7 9	2 9 8 × 9

Determine the place value of the underlined digit in each number.

17. $8.5<u>9</u> _____ $<u>5</u>4.29 _____

Add or subtract. Watch the signs.

	a	b	c	d	e
18.	8 1¢ −2 7¢	6 4¢ +1 8¢	9 5¢ −7 6¢	$4.0 8 −0.5 6	$2 2.4 3 + 9.3 1

Multiply.

19.	$0.2 3 ×4	$1.0 5 ×6	$1.6 4 ×8	$0.8 8 ×2 4	$2.9 3 ×3 2

Complete each of the following.

20.   5 cm = _____ mm   6 km = _____ m

21.   9 kg = _____ g   50 liters = _____ mL

Record the temperature shown by each thermometer.

22.

The temperature is _____ °F.   The temperature is _____ °C.

GO

Complete each of the following.

	a	b
25.	2 ft = _____ in.	3 lb = _____ oz
26.	6 qt = _____ pt	12 gal = _____ qt
27.	5T = _____ lb	3 yd = _____ ft
28.	10 cups = _____ pt	2 days = _____ hours
29.	4 min = _____ sec	6 hours = _____ min

Find the perimeter of each figure.

30.

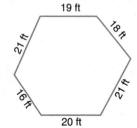

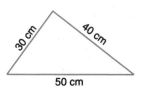

_____ feet

_____ cm

Find the area of each figure.

31.

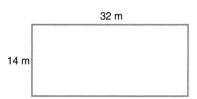

_____ square meters

_____ square inches

Identify each figure.

32.

_____   _____   _____   _____

GO

FINAL TEST Chapters 1-16 (continued)

Compare the figures. Write *congruent* or *not congruent*.

33.   *a*   *b*   *c*

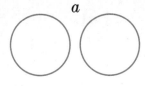   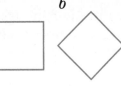

_____   _____   _____

Find the mean of each set of numbers.

            a   *b*

34.   6, 9, 12, 15, 19, 22, 28, 33   45, 73, 59, 26, 54, 19, 55, 74, 97, 88

            _____   _____

Find the median, mode, and range of each set of numbers.

35. 6, 3, 4, 7, 1, 5, 1, 7, 3, 3, 2, 5, 3   12, 16, 13, 16, 18, 15, 12, 11, 16, 14, 19

      median: _____   median: _____

      mode: _____   mode: _____

      range: _____   range: _____

Use the graph to answer each question.

36. How many more hot dogs were sold than hamburgers?

_____ more hot dogs were sold.

37. How many more ice-cream cones were sold than watermelon slices?

_____ more ice-cream cones were sold.

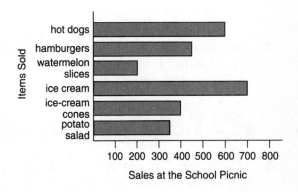

Refer to the wheel to answer the questions.

38. What is the probability of spinning a four?

The probability is _____.

39. What is the probability of spinning a seven?

The probability is _____.

GO

FINAL TEST
Chapters 1-16

Solve each problem.

40. A factory can make 128 hammers in eight hours. The same number of hammers are made each hour. How many hammers do they make each hour?

They make _____ hammers each hour.

41. Jordan earned $490 each week at his job. He also earned $48 each week from a part-time job. How much did Jordan earn in all each week?

Jordan earned $_____ in all each week.

42. A gasoline station sells an average of 847 gallons of gasoline every day. How many gallons will be sold in 365 days?

_____ gallons will be sold.

43. It takes 6 minutes to make one card. How many cards can be made in 8 hours (480 minutes)?

_____ cards can be made.

44. Meghan drove 18,322 miles last year. Ellen drove 9,785 miles last year. How many more miles did Meghan drive than Ellen last year?

Meghan drove _____ more miles than Ellen.

45. There are 1,455 items to be packed. The items are to be packed 8 to a box. How many boxes can be filled? How many items will be left?

_____ boxes will be filled.

_____ items will be left.

40.

41.

42.

43.

44.

45.

CHAPTER 1 CUMULATIVE REVIEW

Work each problem.
Find the correct answer.
Mark the space for the answer.

Part 1 Concepts

1. The sum of 10 and 3 is _____.

A 7
B 13
C 30
D 14

2. The difference between 9 and 5 is ____.

F 14
G 5
H 15
J 4

Part 2 Computation

Solve.

3. 45
   +1

A 46
B 45
C 44
D 56

4.   7
   +32

F 37
G 49
H 39
J 35

5. 89
   −3

A 92
B 76
C 86
D 82

6. 42
   +43

F 85
G 86
H 81
J 75

7. 35
   −21

A 56
B 14
C 4
D 66

Part 3 Applications

8. Paul had 37 dollars. He bought a shirt for 15 dollars. How much money did he have left?

F 12 dollars
G 42 dollars
H 22 dollars
J 24 dollars

9. Marta has 41 marbles. Julie has 56 marbles. How many marbles do the two girls have in all?

A 90   C 15
B 97   D 95

10. Kenny can kick a ball 14 yards. Tom can kick the same ball 27 yards. How many yards farther can Tom kick the ball than Kenny?

F 10   H 14
G 13   J 12

CUMULATIVE REVIEW

ANSWER ROW   **1** Ⓐ Ⓑ Ⓒ Ⓓ   **3** Ⓐ Ⓑ Ⓒ Ⓓ   **5** Ⓐ Ⓑ Ⓒ Ⓓ   **7** Ⓐ Ⓑ Ⓒ Ⓓ   **9** Ⓐ Ⓑ Ⓒ Ⓓ
             2 Ⓕ Ⓖ Ⓗ Ⓙ   **4** Ⓕ Ⓖ Ⓗ Ⓙ   **6** Ⓕ Ⓖ Ⓗ Ⓙ   **8** Ⓕ Ⓖ Ⓗ Ⓙ   **10** Ⓕ Ⓖ Ⓗ Ⓙ

CHAPTER 2 CUMULATIVE REVIEW

Work each problem.
Find the correct answer.
Mark the space for the answer.

Part 1 Concepts

1. The sum of 56 and 87 is _____.

  A 133
  B 143
  C 141
  D 149

2. The difference between 582 and 174 is _____.

  F 406
  G 396
  H 408
  J 402

Part 2 Computation

3.
```
  4 6
+ 3 8
```
  A 74
  B 72
  C 82
  D 84

4.
```
7 0 3
+ 4 9
```
  F 749
  G 742
  H 752
  J 754

5.
```
  6 1 8
+ 5 9 3
```
  A 1,211
  B 1,121
  C 1,210
  D 1,201

6.
```
  3 5 6
    9 4
+ 8 3
```
  F 1,379
  G 533
  H 1,253
  J 433

7.
```
    6 2
    7 5
+ 2 1 9
```
  A 356
  B 896
  C 1,031
  D 286

Part 3 Applications

8. Edie drove 36 miles one day and 82 miles the next day. How many miles did she drive in all?

  F 108   H 118
  G 98   J 138

9. Qadry earned 54 dollars mowing lawns in June and 88 dollars mowing lawns in July. How many dollars did he earn in all?

  A 142 dollars
  B 152 dollars
  C 144 dollars
  D 150 dollars

10. Ling Yi put up 484 feet of fencing for his chickens and 398 feet of fencing around his garden. How many feet of fencing did he put up in all?

  F 900   H 912
  G 894   J 882

ANSWER ROW **1** Ⓐ Ⓑ Ⓒ Ⓓ **3** Ⓐ Ⓑ Ⓒ Ⓓ **5** Ⓐ Ⓑ Ⓒ Ⓓ **7** Ⓐ Ⓑ Ⓒ Ⓓ **9** Ⓐ Ⓑ Ⓒ Ⓓ
           2 Ⓕ Ⓖ Ⓗ Ⓙ **4** Ⓕ Ⓖ Ⓗ Ⓙ **6** Ⓕ Ⓖ Ⓗ Ⓙ **8** Ⓕ Ⓖ Ⓗ Ⓙ **10** Ⓕ Ⓖ Ⓗ Ⓙ

CHAPTER 3 CUMULATIVE REVIEW

Work each problem.
Find the correct answer.
Mark the space for the answer.

Part 1 Concepts

1. Which of these is the best estimate of 174 + 582?

A 600

B 800

C 700

D 900

2. The difference between 47 and 29 is _____.

F 13

G 12

H 15

J 18

Part 2 Computation

3. 7 8 4
  +1 3

A 797

B 766

C 671

D 780

4.   7 8 2
  +1 2 6

F 664

G 907

H 908

J 904

5. 3 8 9
  −3 9

A 349

B 448

C 350

D 450

6. 1 2 4 2
  +5 4 3

F 1,785

G 1,686

H 1,781

J 1,885

7. 3 4 6 7 5
  −5 6 2 1

A 29,051

B 29,054

C 29,154

D 28,055

Part 3 Applications

8. Jake bought a motorbike for 3,588 dollars. How much did he pay to the nearest hundred dollars?

F 3,500 dollars

G 4,000 dollars

H 3,600 dollars

J 3,590 dollars

9. Cheryl has a farm with 186 acres in one section and 13 acres in another. About how many acres are there in all?

A 190   C 100

B 200   D 193

10. The Sears Tower is 1,454 feet tall. It has a 346-foot TV antenna on top. What is the height of the building and the antenna together?

F 1,729 feet   H 1,100 feet

G 1,800 feet   J 1,801 feet

STOP

CUMULATIVE REVIEW

ANSWER ROW **1** Ⓐ Ⓑ Ⓒ Ⓓ **3** Ⓐ Ⓑ Ⓒ Ⓓ **5** Ⓐ Ⓑ Ⓒ Ⓓ **7** Ⓐ Ⓑ Ⓒ Ⓓ **9** Ⓐ Ⓑ Ⓒ Ⓓ

2 Ⓕ Ⓖ Ⓗ Ⓙ **4** Ⓕ Ⓖ Ⓗ Ⓙ **6** Ⓕ Ⓖ Ⓗ Ⓙ **8** Ⓕ Ⓖ Ⓗ Ⓙ **10** Ⓕ Ⓖ Ⓗ Ⓙ

NAME _____

Work each problem.
Find the correct answer.
Mark the space for the answer.

Part 1 Concepts

1. The difference between 283 and 135 is closest to _____.

   A 130
   B 140
   C 150
   D 160

2. Which of these is the best estimate of 8,684 − 3,582?

   F 4,000
   G 5,000
   H 6,000
   J 7,000

Part 2 Computation

3.
$$\begin{array}{r} 5736 \\ +2743 \\ \hline \end{array}$$
   A 8,297
   B 8,173
   C 8,479
   D 7,979

4.
$$\begin{array}{r} 6082 \\ -567 \\ \hline \end{array}$$
   F 5,515
   G 5,483
   H 5,538
   J 5,155

5.
$$\begin{array}{r} 32 \\ \times 4 \\ \hline \end{array}$$
   A 126
   B 138
   C 128
   D 125

6.
$$\begin{array}{r} 5473 \\ 1968 \\ +2396 \\ \hline \end{array}$$
   F 9,927
   G 9,837
   H 8,947
   J 9,037

7.
$$\begin{array}{r} 476 \\ \times 6 \\ \hline \end{array}$$
   A 2,646
   B 3,000
   C 2,776
   D 2,856

Part 3 Applications

8. The pet shop has five large fish tanks. Each tank has 86 fish. How many fish does the pet shop have in all?

   F 430   H 424
   G 540   J 440

9. A trucker drove 3,847 miles in April and 6,935 miles in May. How many more miles did he drive in May than in April?

   A 3,251   C 3,088
   B 10,809   D 4,161

10. Alice earns 328 dollars each week. How much money will she earn in four weeks?

   F 1,213 dollars
   G 1,312 dollars
   H 1,282 dollars
   J 332 dollars

STOP

ANSWER ROW   **1** Ⓐ Ⓑ Ⓒ Ⓓ   **3** Ⓐ Ⓑ Ⓒ Ⓓ   **5** Ⓐ Ⓑ Ⓒ Ⓓ   **7** Ⓐ Ⓑ Ⓒ Ⓓ   **9** Ⓐ Ⓑ Ⓒ Ⓓ
                 2 Ⓕ Ⓖ Ⓗ Ⓙ   **4** Ⓕ Ⓖ Ⓗ Ⓙ   **6** Ⓕ Ⓖ Ⓗ Ⓙ   **8** Ⓕ Ⓖ Ⓗ Ⓙ   **10** Ⓕ Ⓖ Ⓗ Ⓙ

Work each problem.
Find the correct answer.
Mark the space for the answer.

Part 1 Concepts

1. Which problem would you use to find the product of 27 and 376?

   A 27 ÷ 376
   B 30 × 376
   C 376 ÷ 27
   D 27 × 376

2. Which of these is the product of 58 × 779?

   F 45,182
   G 43,102
   H 41,582
   J 1,300

Part 2 Computation

3. $\begin{array}{r} 73 \\ \times 46 \end{array}$
   A 3,308
   B 3,078
   C 3,248
   D 3,358

4. $\begin{array}{r} 682 \\ \times 16 \end{array}$
   F 1,912
   G 12,912
   H 10,912
   J 10,812

5. $\begin{array}{r} 7526 \\ -4971 \end{array}$
   A 2,455
   B 2,745
   C 2,555
   D 2,645

6. $\begin{array}{r} 6108 \\ 4830 \\ +1492 \end{array}$
   F 11,430
   G 12,430
   H 12,340
   J 12,530

7. $\begin{array}{r} 476 \\ \times 35 \end{array}$
   A 16,660
   B 16,606
   C 16,600
   D 16,560

Part 3 Applications

8. Kordell threw a football 78 yards. There are 36 inches in a yard. How many inches did Kordell throw the football?

   F 3,200   H 2,986
   G 2,808   J 3,042

9. The nursery raised 675 rose bushes. Each bush had 12 roses. How many roses did they raise in all?

   A 890   C 8,092
   B 8,142   D 8,100

10. At the zoo, each of the 12 giraffes eats 150 pounds of hay per day. How many pounds of hay does the zoo need each day for the giraffes?

    F 1,800   H 1,650
    G 1,600   J 1,750

CUMULATIVE REVIEW

ANSWER ROW   **1** Ⓐ Ⓑ Ⓒ Ⓓ   **3** Ⓐ Ⓑ Ⓒ Ⓓ   **5** Ⓐ Ⓑ Ⓒ Ⓓ   **7** Ⓐ Ⓑ Ⓒ Ⓓ   **9** Ⓐ Ⓑ Ⓒ Ⓓ
                 2 Ⓕ Ⓖ Ⓗ Ⓙ   **4** Ⓕ Ⓖ Ⓗ Ⓙ   **6** Ⓕ Ⓖ Ⓗ Ⓙ   **8** Ⓕ Ⓖ Ⓗ Ⓙ   **10** Ⓕ Ⓖ Ⓗ Ⓙ

NAME _____

Work each problem.
Find the correct answer.
Mark the space for the answer.

Part 1 Concepts

1. Which problem would you use to find the product of 16 and 4,576?

A 4576 − 16
B 4576 ÷ 16
c 4576 × 16
D 4576 + 16

2. Which of these would you use to find the difference between 6,831 and 28?

F 28 + 6831
G 6831 + 28
H 28 ÷ 6831
J 6831 − 28

Part 2 Computation

3. 3 5 6 2
　　×8

A 12,450
B 28,078
c 23,248
D 28,496

4. 9 4 5
　×2 5

F 23,625
G 23,025
H 21,625
J 21,025

5.　4 8 7 6
　+8 2 2 7

A 12,103
B 13,103
c 13,093
D 12,093

6. 7 9 3 5 2
　−4 7 2 9

F 74,623
G 75,623
H 75,723
J 74,623

7.　3 3 9
　×2 5 1

A 85,089
B 84,079
c 85,709
D 84,089

Part 3 Applications

8. A store sells nails by the pound. If there are 562 nails in each pound, how many nails are in 25 pounds?

F 12,950　　H 14,050
G 11,950　　J 13,950

9. Heidi rode her bicycle 12 miles. There are 5,280 feet in one mile. How many feet did Heidi ride?

A 63,360　　c 62,360
B 63,280　　D 62,280

10. One truck can haul 45 tons of materials. There are 2,000 pounds in each ton. How many pounds of materials can the truck haul?

F 8,000　　H 80,000
G 9,000　　J 90,000

STOP

ANSWER ROW　**1** Ⓐ Ⓑ Ⓒ Ⓓ　**3** Ⓐ Ⓑ Ⓒ Ⓓ　**5** Ⓐ Ⓑ Ⓒ Ⓓ　**7** Ⓐ Ⓑ Ⓒ Ⓓ　**9** Ⓐ Ⓑ Ⓒ Ⓓ
　　　　　　　2 Ⓕ Ⓖ Ⓗ Ⓙ　**4** Ⓕ Ⓖ Ⓗ Ⓙ　**6** Ⓕ Ⓖ Ⓗ Ⓙ　**8** Ⓕ Ⓖ Ⓗ Ⓙ　**10** Ⓕ Ⓖ Ⓗ Ⓙ

CHAPTER 7 CUMULATIVE REVIEW

Work each problem.
Find the correct answer.
Mark the space for the answer.

Part 1 Concepts

1. Which of the following dollar amounts has a 7 in the hundredths place?

  A $734.92

  B $672.17

  C $471.08

  D $356.68

2. Which of these shows 6 dollars and 28 cents?

  F $62.80

  G $0.628

  H $6.028

  J $6.28

Part 2 Computation

3.
$$\begin{array}{r} 3\,6¢ \\ +4\,7¢ \\ \hline \end{array}$$

  A 83¢

  B $87

  C 73¢

  D 84¢

4.
$$\begin{array}{r} \$1.7\,5 \\ -0.9\,4 \\ \hline \end{array}$$

  F $08.1

  G $081

  H $0.81

  J $8.10

5.
$$\begin{array}{r} \$7.9\,3 \\ +4.2\,7 \\ \hline \end{array}$$

  A $122.0

  B $11.10

  C $11.20

  D $12.20

6.
$$\begin{array}{r} 5\,3\,8 \\ \times 6\,4 \\ \hline \end{array}$$

  F 32,882

  G 33,962

  H 34,432

  J 30,462

7.
$$\begin{array}{r} 6\,2\,3\,1 \\ \times 8 \\ \hline \end{array}$$

  A 48,848

  B 48,838

  C 49,848

  D 49,838

Part 3 Applications

8. Tomatoes cost $2.09 per pound. What is the cost of 5 pounds of tomatoes?

  F $1045   H $10.45

  G $9.45   J $104.50

9. Automobile tires are on sale for $86.25 each. What is the cost of four tires?

  A $34.50   C $445.00

  B $345.00   D $355.50

10. Aaron bought a CD for $15.95. He gave the clerk a twenty-dollar bill. How much change did he receive?

  F $4.05   H $5.05

  G $5.15   J $4.15

STOP

CUMULATIVE REVIEW

ANSWER ROW **1** Ⓐ Ⓑ Ⓒ Ⓓ **3** Ⓐ Ⓑ Ⓒ Ⓓ **5** Ⓐ Ⓑ Ⓒ Ⓓ **7** Ⓐ Ⓑ Ⓒ Ⓓ **9** Ⓐ Ⓑ Ⓒ Ⓓ
           2 Ⓕ Ⓖ Ⓗ Ⓙ **4** Ⓕ Ⓖ Ⓗ Ⓙ **6** Ⓕ Ⓖ Ⓗ Ⓙ **8** Ⓕ Ⓖ Ⓗ Ⓙ **10** Ⓕ Ⓖ Ⓗ Ⓙ

CHAPTER 8 CUMULATIVE REVIEW

Work each problem.
Find the correct answer.
Mark the space for the answer.

Part 1 Concepts

1. Which of the following division facts has a quotient of 6?

   A 4)24   C 5)25

   B 6)30   D 7)49

2. Which division fact completes the sentence correctly?

   $$\begin{array}{r} 7 \\ \times 8 \\ \hline 56 \end{array}$$ so _____

   F $\begin{array}{r} 7 \\ 64)\overline{8} \end{array}$ H $\begin{array}{r} 7 \\ 8)\overline{56} \end{array}$

   G $\begin{array}{r} 8 \\ 7)\overline{64} \end{array}$ J $\begin{array}{r} 8 \\ 64)\overline{7} \end{array}$

Part 2 Computation

3. 4)3 2

   A 7
   B 6
   C 9
   D 8

4. 7)2 8

   F 5
   G 4
   H 3
   J 6

5. 6)0

   A 1
   B 0
   C 3
   D 6

6. $\begin{array}{r} 8\,4\,9\,2 \\ \times 1\,8 \\ \hline \end{array}$

   F 16,996
   G 152,856
   H 149,586
   J 151,786

7. $\begin{array}{r} \$8\,3\,0.7\,1 \\ -\,8\,7.9\,7 \\ \hline \end{array}$

   A $742.74
   B $733.74
   C $740.74
   D $755.74

Part 3 Applications

8. Soledad baked 24 cookies. They are divided evenly among eight people. How many cookies will each person get?

   F 5   H 4
   G 3   J 2

9. Ed bought 35 bones for his dogs. He has five dogs and shares the bones evenly. How many bones will each dog get?

   A 5   C 8
   B 7   D 6

10. Sylvia had a piece of ribbon that was 64 inches long. She cut the ribbon into eight equal pieces. How many pieces does she have now?

   F 5   H 8
   G 9   J 7

STOP

ANSWER ROW 1 (A)(B)(C)(D) 3 (A)(B)(C)(D) 5 (A)(B)(C)(D) 7 (A)(B)(C)(D) 9 (A)(B)(C)(D)
            2 (F)(G)(H)(J) 4 (F)(G)(H)(J) 6 (F)(G)(H)(J) 8 (F)(G)(H)(J) 10 (F)(G)(H)(J)

NAME _____

Work each problem.
Find the correct answer.
Mark the space for the answer.

Part 1 Concepts

1. Which of the following division problems has a remainder of 3?

A 3)‾16‾   C 7)‾23‾

B 6)‾35‾   D 5)‾48‾

2. Which pair of multiplication problems would help you divide 86 by 4?

F $4 \times 20 = 80$ H $4 \times 10 = 40$
   $4 \times 10 = 40$ $4 \times 30 = 120$

G $4 \times 20 = 80$ J $3 \times 20 = 60$
   $4 \times 30 = 120$ $4 \times 20 = 80$

Part 2 Computation

3. 4)‾8 7‾

A 21 r3
B 23
C 21
D 22 r2

4. 7)‾3 4 5‾

F 49
G 50
H 49 r2
J 48 r6

5. 6)‾2 5 2‾

A 42 r3
B 41 r5
C 40
D 42

6. 789
    ×321

F 210,759
G 253,269
H 249,269
J 254,659

7. $9 4 5.0 8
    +3 6 5.2 6

A $1,290.54
B $1,309.64
C $1,310.34
D $1,312.54

Part 3 Applications

8. The school placed eight benches around the soccer field. The benches were 96 feet long in all. How many feet long was each bench?

F 10   H 13
G 9 r7   J 12

9. LaShawna jogged 105 kilometers last week. If she jogged the same number of kilometers each day, how many kilometers did she jog on Monday?

A 15   C 14 r2
B 13   D 18

10. One 504-mile bicycle race is divided into eight equal legs. How many miles is each leg?

F 80   H 60 r24
G 63   J 65

CUMULATIVE REVIEW

ANSWER ROW **1** Ⓐ Ⓑ Ⓒ Ⓓ   **3** Ⓐ Ⓑ Ⓒ Ⓓ   **5** Ⓐ Ⓑ Ⓒ Ⓓ   **7** Ⓐ Ⓑ Ⓒ Ⓓ   **9** Ⓐ Ⓑ Ⓒ Ⓓ
            2 Ⓕ Ⓖ Ⓗ Ⓙ   **4** Ⓕ Ⓖ Ⓗ Ⓙ   **6** Ⓕ Ⓖ Ⓗ Ⓙ   **8** Ⓕ Ⓖ Ⓗ Ⓙ   **10** Ⓕ Ⓖ Ⓗ Ⓙ

NAME _____

Work each problem.
Find the correct answer.
Mark the space for the answer.

Part 1 Concepts

1. Which of the following division problems has a quotient with a digit in the thousands place?

A $4\overline{)1678}$ C $3\overline{)5821}$

B $8\overline{)3510}$ D $7\overline{)4811}$

2. Which of the following division problems will have a remainder?

F $2\overline{)448}$ H $2\overline{)1897}$

G $2\overline{)510}$ J $2\overline{)6286}$

Part 2 Computation

3. $7\overline{)583}$

A 93 r5
B 83 r2
C 84 r1
D 79 r6

4. $5\overline{)835}$

F 167
G 165 r4
H 167 r3
J 166 r5

5. $9\overline{)5104}$

A 577 r3
B 569 r4
C 568 r7
D 567 r1

6.
$$\begin{array}{r} 6\,3\,4\,2 \\ 2\,7\,4\,5 \\ +\,6\,7\,4 \\ \hline \end{array}$$

F 8,651
G 9,651
H 9,761
J 8,661

7.
$$\begin{array}{r} \$8\,6.0\,2 \\ -\,3\,7.5\,7 \\ \hline \end{array}$$

A $48.45
B $58.55
C $59.55
D $49.55

Part 3 Applications

8. A 375-dollar prize will be evenly distributed among five winners. How much money will each winner receive?

F 73 dollars   H 75 dollars
G 90 dollars   J 80 dollars

9. There were 4,104 concert tickets sold in one season. The same number of people attended each of nine concerts. How many people attended each concert?

A 456   C 455 r3
B 457   D 467

10. Brenda packs 8 donuts in each box. She has to pack 1,352 donuts. How many boxes will she fill? How many donuts will be left?

F 169 boxes, 3 left
G 169 boxes, 0 left
H 172 boxes, 5 left
J 168 boxes, 4 left

STOP

ANSWER ROW   **1** Ⓐ Ⓑ Ⓒ Ⓓ   **3** Ⓐ Ⓑ Ⓒ Ⓓ   **5** Ⓐ Ⓑ Ⓒ Ⓓ   **7** Ⓐ Ⓑ Ⓒ Ⓓ   **9** Ⓐ Ⓑ Ⓒ Ⓓ
             2 Ⓕ Ⓖ Ⓗ Ⓙ   **4** Ⓕ Ⓖ Ⓗ Ⓙ   **6** Ⓕ Ⓖ Ⓗ Ⓙ   **8** Ⓕ Ⓖ Ⓗ Ⓙ   **10** Ⓕ Ⓖ Ⓗ Ⓙ

NAME _____

Work each problem.
Find the correct answer.
Mark the space for the answer.

Part 1 ·Concepts

1. Which of the following is the best estimate for $9\overline{)462}$?

A 35   C 60
B 58   D 50

2. When checking a division problem using multiplication, the dividend should be the same as the:

F sum   H divisor
G product   J quotient

3. Round 48,392 to the nearest hundred.

A 48,400   C 48,000
B 50,000   D 48,390

Part 2 Computation

4. $8\overline{)9347}$

F 1,168 r3
G 1,068 r3
H 1,178 r1
J 1,168

5. 6371
   ×28

A 168,378
B 174,288
C 178,388
D 177,388

6. $9\overline{)5104}$

F 566 r4
G 567 r1
H 557 r1
J 568

7. 74825
   −9372

A 75,533
B 65,553
C 65,453
D 75,453

Part 3 Applications

8. If Carlo earns $3,695 per month, how much money will he earn in nine months?

F $36,000   H $32,155
G $33,255   J $33,000

9. The first 750 people at the ball game received a free cap. The caps cost the team $3.67 each. How much did the caps cost the team?

A $2,643.75   C $2,800.00
B $2,755.00   D $2,752.50

10. There are 288 fourth-grade students at Laurel School. There are nine fourth-grade homerooms. If every homeroom has the same number of students, how many students are in each homeroom?

F 32
G 30 r5
H 29 r8
J 31 r6

ANSWER ROW **1** Ⓐ Ⓑ Ⓒ Ⓓ   **3** Ⓐ Ⓑ Ⓒ Ⓓ   **5** Ⓐ Ⓑ Ⓒ Ⓓ   **7** Ⓐ Ⓑ Ⓒ Ⓓ   **9** Ⓐ Ⓑ Ⓒ Ⓓ
                2 Ⓕ Ⓖ Ⓗ Ⓙ   **4** Ⓕ Ⓖ Ⓗ Ⓙ   **6** Ⓕ Ⓖ Ⓗ Ⓙ   **8** Ⓕ Ⓖ Ⓗ Ⓙ   **10** Ⓕ Ⓖ Ⓗ Ⓙ

CHAPTER 12 CUMULATIVE REVIEW

Work each problem.
Find the correct answer.
Mark the space for the answer.

Part 1 Concepts

1. Which shows two equivalent fractions?

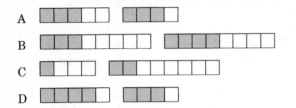

2. Which fraction describes part of something divided into five equal sections?

F $\frac{5}{9}$ H $\frac{2}{3}$

G $\frac{3}{5}$ J $\frac{1}{4}$

3. Round 67,294 to the nearest thousand.

A 70,000   C 67,300

B 67,000   D 67,290

Part 2 Computation

4. $\frac{6}{8} - \frac{2}{8}$

F $\frac{8}{8}$ H $\frac{4}{0}$

G $\frac{4}{8}$ J 1

5. $\frac{1}{6} + \frac{2}{6}$

A $\frac{3}{12}$ C $\frac{3}{6}$

B $\frac{1}{12}$ D 3

6. $\begin{array}{r} \$0.6\,8 \\ \times 3\,6 \\ \hline \end{array}$

F $24.48

G $23.58

H $25.48

J $36.00

7. $\begin{array}{r} 4\,8\,5\,7\,4 \\ +8\,9\,6\,7 \\ \hline \end{array}$

A 57,541

B 47,641

C 56,741

D 57,341

Part 3 Applications

8. If Hank reads 87 pages each week, how many pages will he read in a year (52 weeks)?

F 4,524   H 4,154

G 4,434   J 4,554

9. Lenore read $\frac{2}{5}$ of a book yesterday. Today she read $\frac{1}{5}$ of the book. What fractional part of the book has she read?

A $\frac{2}{5}$ C $\frac{3}{5}$

B $\frac{3}{10}$ D 3

10. The Johnsons ordered a pizza. Mike ate two pieces, Ashton ate three pieces, Colin ate four pieces, Molly ate two pieces, and the dog ate the last piece. What fraction of the pizza did Colin eat?

F $\frac{4}{12}$ H $\frac{4}{10}$

G $\frac{4}{4}$ J $\frac{4}{1}$

ANSWER ROW   **1** Ⓐ Ⓑ Ⓒ Ⓓ   **3** Ⓐ Ⓑ Ⓒ Ⓓ   **5** Ⓐ Ⓑ Ⓒ Ⓓ   **7** Ⓐ Ⓑ Ⓒ Ⓓ   **9** Ⓐ Ⓑ Ⓒ Ⓓ

2 Ⓕ Ⓖ Ⓗ Ⓙ   **4** Ⓕ Ⓖ Ⓗ Ⓙ   **6** Ⓕ Ⓖ Ⓗ Ⓙ   **8** Ⓕ Ⓖ Ⓗ Ⓙ   **10** Ⓕ Ⓖ Ⓗ Ⓙ

CHAPTER 13 CUMULATIVE REVIEW

Work each problem.
Find the correct answer.
Mark the space for the answer.

Part 1 Concepts

1. Angles can be named by:
   A how sharp they are.
   B the position of the two sides.
   C their size.
   D their color.

2. Two figures are congruent if they have the same _____.
   F size   H area
   G shape   J size and shape

Part 2 Computation

3. 6)3 8 5 7
   A 632 r1   C 641 r11
   B 599 r5   D 642 r5

4.   7 7 5
     ×4 8 2
   F 363,655   H 400,000
   G 373,550   J 372,560

5. 5)8 3 2 6
   A 1,675 r2   C 1,665 r1
   B 1,659 r4   D 1,664 r6

6. $\frac{7}{10} - \frac{4}{10}$
   F $\frac{3}{0}$ H $\frac{3}{20}$
   G $\frac{11}{10}$ J $\frac{3}{10}$

7. 8 3 6 5 7
   +9 9 4 7
   A 93,604
   B 94,704
   C 73,710
   D 74,810

Part 3 Applications

8. Jenny drew a hexagon in art class. Which of these did she draw?

   F   H

   G   J

9. John Grayson Oliver Roberts wants to use one of his initials to describe symmetry to a friend. Which initial should he use?
   A J   C O
   B G   D R

10. Main Street crosses Maple Avenue. The two roads are perpendicular to each other. How many right angles does the crossing form?
    F 2   H 4
    G 3   J 5

STOP

CUMULATIVE REVIEW

ANSWER ROW   1 Ⓐ Ⓑ Ⓒ Ⓓ   3 Ⓐ Ⓑ Ⓒ Ⓓ   5 Ⓐ Ⓑ Ⓒ Ⓓ   7 Ⓐ Ⓑ Ⓒ Ⓓ   9 Ⓐ Ⓑ Ⓒ Ⓓ
              2 Ⓕ Ⓖ Ⓗ Ⓙ   4 Ⓕ Ⓖ Ⓗ Ⓙ   6 Ⓕ Ⓖ Ⓗ Ⓙ   8 Ⓕ Ⓖ Ⓗ Ⓙ   10 Ⓕ Ⓖ Ⓗ Ⓙ

CHAPTER 14 CUMULATIVE REVIEW

NAME _____

Work each problem.
Find the correct answer.
Mark the space for the answer.

Part 1 Concepts

1. Which of these operations or pairs of operations could be used to find perimeter?

A subtraction
B division
C addition and multiplication
D multiplication and subtraction

2. If you wanted to find the number of millimeters in 13 centimeters, which of these operations would you use?

F $13 \times 1,000$ H $13 - 10$
G $10\overline{)13}$ J 13×10

Part 2 Computation

3. $\begin{array}{r} 3857 \\ +9368 \end{array}$

A 12,235
B 13,225
C 13,125
D 13,335

4. $\begin{array}{r} 1000 \\ \times\ 98 \end{array}$

F 9,800
G 98,000
H 980,000
J 980

5. $4\overline{)9674}$

A 2,418 r2
B 2,617 r3
C 2,328 r1
D 2,500

6. $\frac{14}{15} - \frac{6}{15}$

F $\frac{20}{15}$ H $\frac{12}{15}$
G $\frac{8}{15}$ J 8

7. $\begin{array}{r} 10130 \\ -7845 \end{array}$

A 17,975
B 3,395
C 2,375
D 2,285

Part 3 Applications

8. JoAnne jogs 3 kilometers every morning. How many meters does she jog?

F 300,000   H 30,000
G 3,000   J 300

9. Jorge brought 5,000 milliliters of fruit punch to the school picnic. How many liters of punch did he bring?

A 50   C 5
B 500   D 50,000

10. A courtyard in a hotel measures 35 meters by 28 meters. What is the area of the courtyard?

F 980 square meters
G 1,000 square meters
H 126 square meters
J 63 square meters

STOP

ANSWER ROW **1** Ⓐ Ⓑ Ⓒ Ⓓ **3** Ⓐ Ⓑ Ⓒ Ⓓ **5** Ⓐ Ⓑ Ⓒ Ⓓ **7** Ⓐ Ⓑ Ⓒ Ⓓ **9** Ⓐ Ⓑ Ⓒ Ⓓ
2 Ⓕ Ⓖ Ⓗ Ⓙ **4** Ⓕ Ⓖ Ⓗ Ⓙ **6** Ⓕ Ⓖ Ⓗ Ⓙ **8** Ⓕ Ⓖ Ⓗ Ⓙ **10** Ⓕ Ⓖ Ⓗ Ⓙ

Work each problem.
Find the correct answer.
Mark the space for the answer.

Part 1 Concepts

1. Which of these operations or pairs of operations could be used to find area?

   A subtraction

   B multiplication

   C addition and multiplication

   D division

2. If you wanted to find the number of inches in 34 feet, which operation would you use?

   F $34 \times 1,000$ H 12×34

   G $12\overline{)34}$ J 34×100

3. Which fraction describes part of a figure divided into eight equal sections?

   A $\frac{6}{8}$ C $\frac{5}{6}$

   B $\frac{8}{12}$ D $\frac{8}{1}$

4. To find the number of pounds in 29 ounces, which equation would you use?

   F 29×16 H $2\overline{)16}$

   G $16\overline{)29}$ J 16×2

5. Which of these division facts completes the sentence correctly?

   $\begin{array}{r} 9 \\ \times 8 \\ \hline 72 \end{array}$ so _____

   A $72\overline{)8}^{\,9}$ C $8\overline{)72}^{\,9}$

   B $9\overline{)81}^{\,8}$ D $8\overline{)17}^{\,9}$

Part 2 Computation

6. $\begin{array}{r} 8\,5\,1\,2 \\ -3\,7\,9\,5 \\ \hline \end{array}$

   F 4,717

   G 5,617

   H 5,717

   J 4,817

7. $\begin{array}{r} 3\,8\,5\,7 \\ \times 6\,0 \\ \hline \end{array}$

   A 232,450

   B 232,150

   C 230,750

   D 231,420

8. $\frac{6}{13} + \frac{5}{13}$

   F $\frac{11}{26}$

   G 1

   H $\frac{1}{13}$

   J $\frac{11}{13}$

9. $8\overline{)7\,7\,7\,7}$

   A 972 r1

   B 1,000

   C 961 r7

   D 969 r2

10. Estimate the product:

   $\begin{array}{r} 5\,8\,8 \\ \times 3\,2 \\ \hline \end{array}$

   F 15,000

   G 20,000

   H 18,000

   J 24,000

CUMULATIVE REVIEW

ANSWER ROW   **1** Ⓐ Ⓑ Ⓒ Ⓓ   **3** Ⓐ Ⓑ Ⓒ Ⓓ   **5** Ⓐ Ⓑ Ⓒ Ⓓ   **7** Ⓐ Ⓑ Ⓒ Ⓓ   **9** Ⓐ Ⓑ Ⓒ Ⓓ

   2 Ⓕ Ⓖ Ⓗ Ⓙ   **4** Ⓕ Ⓖ Ⓗ Ⓙ   **6** Ⓕ Ⓖ Ⓗ Ⓙ   **8** Ⓕ Ⓖ Ⓗ Ⓙ   **10** Ⓕ Ⓖ Ⓗ Ⓙ

11.
    5 8 6 7
  + 9 3 8 6

   A 14,263
   B 15,353
   C 15,163
   D 15,253

12. 6 qt = _____ pt

   F 6   H 16
   G 1   J 12

Part 3 Applications

13. Peter jogs 3 miles every morning. There are 1,760 yards in one mile. How many yards does Peter jog every morning?

   A 21,120   C 7,040
   B 3,520   D 5,280

14. A punch recipe calls for 3 quarts of soda. Jeanine only has a pint measure. How many pints of soda should she use?

   F 8   H 4
   G 6   J 2

15. A rectangular swimming pool measures 62 feet by 35 feet. What is the area of the swimming pool?

   A 2,270 square feet
   B 194 feet
   C 2,170 square feet
   D 744 inches by 420 inches

16. Yesterday Adam had a temperature of 104°F. Today his temperature is closer to normal at 99°F. How many degrees did his temperature drop?

   F 12°F   H 5°F
   G 15°F   J 10°C

17. Sue withdrew $1,259 from her savings account. Now she has $3,255 left in the account. How much money did she have in the account before the withdrawal?

   A $1,996   C $5,413
   B $2,514   D $4,514

18. A gardener built rectangular fencing around his yard. The fencing measured 36 yards by 48 yards. What is the perimeter of the fencing?

   F 168 yards   H 1,200 yards
   G 1,728 yards   J 178 yards

19. If Tom earns $2,865 per month, how much money will he earn in ten months?

   A $28,650   C $2,875
   B $3,865   D $20,875

20. Bill bought 774 feet of sturdy rope to make a tightrope for a circus. How many yards of rope does he have?

   F 775 yards   H 258 yards
   G 744 yards   J 285 yards

STOP

ANSWER ROW **11** Ⓐ Ⓑ Ⓒ Ⓓ **13** Ⓐ Ⓑ Ⓒ Ⓓ **15** Ⓐ Ⓑ Ⓒ Ⓓ **17** Ⓐ Ⓑ Ⓒ Ⓓ **19** Ⓐ Ⓑ Ⓒ Ⓓ
               12 Ⓕ Ⓖ Ⓗ Ⓙ **14** Ⓕ Ⓖ Ⓗ Ⓙ **16** Ⓕ Ⓖ Ⓗ Ⓙ **18** Ⓕ Ⓖ Ⓗ Ⓙ **20** Ⓕ Ⓖ Ⓗ Ⓙ

CHAPTER 16 CUMULATIVE REVIEW

Work each problem.
Find the correct answer.
Mark the space for the answer.

Part 1 Concepts

1. What type of graph is used to show change over time?

A pyramid graph
B bar graph
C photo graph
D line graph

2. What is the range of a set of numbers?

F difference between the greatest and the least
G mode
H average
J median

3. Which of the following is the best description of the mode of a set of numbers?

A the average
B the number that occurs most often
C the middle number
D the greatest number

4. When tossing a six-sided cube, how many possible outcomes are there?

F 1   H 6
G 4   J $\frac{1}{6}$

5. What is the probability of tossing a one on a six-sided cube?

A 1   C 6
B $\frac{1}{3}$ D $\frac{1}{6}$

Part 2 Computation

6. $\begin{array}{r} 786 \\ \times 522 \\ \hline \end{array}$

F 400,000
G 410,292
H 409,192
J 410,182

7. $\frac{5}{8} - \frac{2}{8}$

A $\frac{7}{8}$
B $\frac{3}{0}$
C $\frac{7}{0}$
D $\frac{3}{8}$

8. $3\overline{)8374}$

F 2,791 r1
G 2,790 r4
H 2,691 r2
J 3,000

9. Find the median of this set of numbers: 13, 22, 13, 16, 19, 23, 11, 27, 32

A 13   C 19
B 22   D 21

10. Find the mean of this set of numbers: 8, 22, 13, 16, 19, 23, 11, 27, 32

F 13   H 22
G 19   J 21

GO ►

ANSWER ROW
1 Ⓐ Ⓑ Ⓒ Ⓓ   3 Ⓐ Ⓑ Ⓒ Ⓓ   5 Ⓐ Ⓑ Ⓒ Ⓓ   7 Ⓐ Ⓑ Ⓒ Ⓓ   9 Ⓐ Ⓑ Ⓒ Ⓓ
2 Ⓕ Ⓖ Ⓗ Ⓙ   4 Ⓕ Ⓖ Ⓗ Ⓙ   6 Ⓕ Ⓖ Ⓗ Ⓙ   8 Ⓕ Ⓖ Ⓗ Ⓙ   10 Ⓕ Ⓖ Ⓗ Ⓙ

Orange Book

CHAPTER 16 CUMULATIVE REVIEW (continued)

11. $\frac{2}{12} + \frac{5}{12}$

A $\frac{6}{12}$

B $\frac{7}{12}$

C $\frac{7}{24}$

D 7

12. $\frac{2}{5} = ?$

F $\frac{4}{5}$

G $\frac{8}{15}$

H $\frac{10}{25}$

J $\frac{4}{10}$

Part 3 Applications

13. What is the probability of spinning a 3 on this wheel?

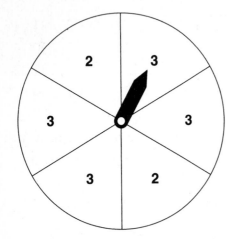

A $\frac{6}{4}$ C $\frac{6}{2}$

B $\frac{4}{6}$ D $\frac{2}{6}$

14. According to the graph, what is the total of blue cars and white cars sold?

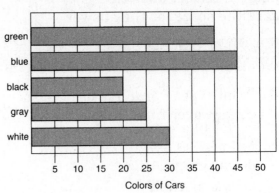

Colors of Cars

F 30   H 70
G 45   J 75

15. The trading club counted the baseball cards each member owned. The six members' total number of cards were 12, 32, 35, 21, 3, and 17. What was the average number of baseball cards owned by club members?

A 20   C 23
B 35   D 22

16. There are three red marbles, five blue marbles, six white marbles, and two pink marbles in a bag. Which fraction represents the probability of choosing a pink marble from the bag?

F $\frac{3}{16}$ H $\frac{2}{16}$

G $\frac{2}{5}$ J $\frac{1}{9}$

STOP

ANSWER ROW **11** Ⓐ Ⓑ Ⓒ Ⓓ   **13** Ⓐ Ⓑ Ⓒ Ⓓ   **15** Ⓐ Ⓑ Ⓒ Ⓓ
            12 Ⓕ Ⓖ Ⓗ Ⓙ   **14** Ⓕ Ⓖ Ⓗ Ⓙ   **16** Ⓕ Ⓖ Ⓗ Ⓙ